The Kaguya Lunar Atlas

Motomaro Shirao • Charles A. Wood

The Kaguya Lunar Atlas

The Moon in High Resolution

 Springer

Authors
Motomaro Shirao
1-3-11, Nishi-Asakusa
Taito-ku
Tokyo 111-0035
Japan

Charles A. Wood
Wheeling Jesuit University
Wheeling, WV 26003
USA
and
Planetary Science Institute
Tucson, AZ 857119
USA

ISBN 978-1-4419-7284-2   e-ISBN 978-1-4419-7285-9
DOI 10.1007/978-1-4419-7285-9
Springer New York Dordrecht Heidelberg London

Jacket design background image: courtesy of JAXA/NHK; Inlet image: courtesy of JAXA.

Back jacket flap image: courtesy of Mitsubishi Heavy Industries, Ltd.

Printed on acid-free paper

Springer is part of Springer Science+Business Media (www.springer.com)

Contents

Part I   Background

1   The Kaguya Mission   3

2   Kaguya's HDTV and Its Imaging   9

3   Images of Earth and Kaguya's Impact on the Moon   23

4   The Lunar Surface   33

Part II   The Atlas

5   Plates 1 to 28   57

6   Plates 29 to 64   87

7   Plates 65 to 100   125

Thumbnail Index   163

Subject Index   169

About the Authors

Motomaro Shirao has been a freelance photographer and science writer since he finished his Master Course in Volcanic Geology at the University of Tokyo in 1980. He has written many articles about the Moon, volcanoes, geology, and geomorphology. His publications include *Graphic Natural History of Volcanoes, Geology and Geomorphology of Japan, Basics of the Moon*, and *Wonderful Landscapes of the World* (all written in Japanese). He is a co-investigator of *Kaguya*'s Terrain Camera and HDTV.

Charles Wood is a senior scientist at the Planetary Science Institute in Arizona and director of the Center for Educational Technologies at Wheeling Jesuit University in West Virginia. He is the author of *The New Moon – A Personal View*, published in 2003; *The Lunar 100 Chart* in 2004; and has written monthly columns about the Moon since 1999 for *Sky & Telescope* magazine. He is the originator of the websites Lunar Photo of the Day and The Moon Wiki. He studies the Moon with the latest spacecraft images and with a small telescope in his backyard.

Acknowledgments

We especially thank JAXA (Japan Aerospace Exploration Agency) and NHK (Japan Broadcasting Corporation). All of the lunar images in this book are, unless otherwise cited, by courtesy of JAXA/NHK. We also especially thank Junichi Yamazaki, Rie Honda, Seiji Mitsuhashi, and Junichi Tachino for cooperation to select the targets of lunar features and help in preparation of this book. We also thank Masaya Kawaguchi, Junichi Haruyama, Manabu Kato, Yoshiaki Ishihara and all the people who contributed to the Kaguya (SELENE) project.

Part I
Background

The *Kaguya* Mission

The Japanese lunar orbiter *Kaguya* was launched on September 14, 2007, by an H-IIA rocket from Tanegashima Space Center (Figure 1.1). *Kaguya* is also known as SELENE, the SELenological and ENgineering Explorer, and was developed by the Japan Aerospace Exploration Agency (JAXA). The nickname *Kaguya*, selected by a popular vote, is the name of a princess in Japanese medieval folklore who came from, and returned to, the Moon. The boxy main orbiter, 2.1 m wide and 4.8 m tall, weighed about 3 t with fuel (Figure 1.2). Two small satellites, Okina and Ouna, each about 50 kg, accompanied the main craft.

After orbiting Earth 2.5 times to achieve the correct position, *Kaguya*'s rockets fired again, directing it to the Moon. On October 4, 2007, the spacecraft was inserted into a large elliptical orbit around the Moon. After mission controllers lowered the spacecraft's altitude on October 18, *Kaguya* finally reached its nominal circular and polar observation orbit of about 100 km above the surface. On the way to nominal orbit, the two small satellites Okina and Ouna were released into elliptical orbits of 100 km perilune (closest point to the Moon), and 2,400 km and 800 km apolune (most distant point from the Moon), respectively. Okina and Ouna always provided a radio beacon and a radio relay when the main spacecraft was behind the Moon. By carefully tracking the motions of all three satellites, dynamicists have created very detailed maps of the Moon's subsurface gravity field.

After checkout of the spacecraft bus system, the four sounder antennas of 15 m length and the 12 m mast for the magnetometer were extended, and the plasma imager was successfully deployed. Following instrument performance tests, which lasted for about 1.5 months, the nominal observation mission of 10 months began on December 21.

The mission's main objectives were to help determine the Moon's origin and evolution and to develop technology for future lunar explorations. The specifications of *Kaguya*'s 13 science instruments are summarized in Table 1.1. The instruments included highly capable cameras, spectrometers, and other detectors. For example, the Terrain Camera, which recorded surface details as small as 10 m wide in stereo, successfully imaged the interior of Shackleton (a deep, perpetually shadowed crater at the lunar south pole) using only the faint light scattered from the crater's sunlit upper rim. *Kaguya*'s laser altimeter measured more than 10 million points of lunar surface and yielded a highly detailed topographic map of the entire Moon. The spacecraft's radar sounder revealed subsurface layers several hundred meters deep under a near-side maria.

At the end of October 2008, the nominal mission was completed successfully, with observational

Figure 1.1 Liftoff (Image courtesy of Mitsubishi Heavy Industries, Ltd.)

coverage of over 95% of the lunar surface. Observations at 100 km altitude were extended for three additional months, to collect additional data.

Low altitude operations at 50 ± 20 km were carried out starting on February 1, 2009. When the relay satellite Okina impacted the lunar far side on February 12, 2009, gravimetric observations were successfully completed. *Kaguya* reduced its altitude again to 10–30 km (perilune) on April 16, 2009.

Finally, at 18:25 UT on June 10, 2009, *Kaguya* impacted the lunar surface before losing attitude control due to fuel exhaustion. The target point was 80.4°E, 65.5°S, on the south-southeastern limb just beyond the terminator on the night side. The 3.9-m Anglo-Australian telescope captured the brief infrared flash of its impact (Figure 3.9b), a celebratory firework marking the spacecraft's successful end of mission.

Figure 1.2 Japan's lunar explorer *Kaguya*'s main orbiter and the two small satellites accompanying it (Image courtesy of JAXA)

Table 1.1 Science instruments of *Kaguya*

Mission instruments	Purpose
X-ray Spectrometer (XRS)	The surface elemental composition (Al, Si, Mg, Fe, etc.) is determined through X-ray fluorescence spectrometry by irradiation of solar X-ray
Gamma-ray Spectrometer (GRS)	The abundance of key elements (U, Th, K, H, etc.) is determined by measuring energy spectra of gamma-rays from the lunar surface with high energy resolution
Multiband Imager (MI)	The mineral distribution is derived from visible and near infrared images of the Moon's surface taken in nine wavelength bands
Spectral Profiler (SP)	The mineral composition of the Moon's surface is obtained by measuring the continuous visible and near infrared spectrum
Terrain Camera (TC)	High-resolution geographical features are acquired by the stereo cameras
Lunar Radar Sounder (LRS)	The subsurface stratification and tectonic features in the shallow part of the lunar crust (a few kilometers) by high-power RF pulses
Laser Altimeter (LALT)	Surface altitudes are precisely measured using high-power laser pulses to make a lunar topography model
Lunar Magnetometer (LMAG)	The magnetization structure on the Moon is acquired by measuring the lunar and the surrounding magnetic field
Charged Particle Spectrometer (CPS)	Alpha rays from the Moon's surface and the abundance of cosmic ray particles are measured
Plasma Energy Angle and Composition Experiment (PACE)	The three dimensional distribution of the low-energy electrons and mass-discriminated low-energy ions around the Moon are measured
Radio Science (RS)	The Moon's ionosphere is detected by measuring the small deviation in the phase of RF signals from "OUNA" (VRAD Satellite)

Table 1.1 (continued)

Mission instruments	Purpose
Upper atmosphere and Plasma Imager (UPI)	Images of the magnetosphere and the ionosphere around the Earth as seen from the Moon are used to study the behavior of the plasma
Four-way Doppler measurements by "OKINA" (Relay Satellite) and main orbiter transponder (RSAT)	Signals from the Main Orbiter in flight on the far side of the Moon are relayed by "OKINA" (Relay Satellite), and the local gravity field data from the far side of the Moon is obtained by measuring the disturbance in the orbit of the Main orbiter using four-way Doppler measurements
Differential VLBI radio source (VRAD)	The gravity field of the Moon is accurately observed by measuring the orbits of the "OKINA" (Relay Satellite) and "OUNA" (VRAD Satellite) using differential VLBI: (Very Long Baseline Interferometry)
High-Definition Television (HDTV)	Taking pictures and movies of the Earth and the Moon with high-definition television cameras

Kaguya's HDTV and Its Imaging

Overview of the HDTV System

In addition to *Kaguya*'s 13 science instruments, the HDTV was unique in being specifically included to engage the public in the excitement of lunar exploration. Although other scientific cameras of *Kaguya*, such as the Terrain Camera and Multiband Imager, were designed to acquire high-resolution images of the area around the nadir, HDTV was optimized for off-nadir observation, giving an astronaut's-eye view of large areas of the lunar surface.

The HDTV system on board *Kaguya* was developed by NHK (Japan Broadcasting Corporation). The system was assembled from consumer broadcast equipment that was modified for the space environment. It consisted of a camera unit and a data processing unit (Figure 2.1). The HDTV system weighed 16.5 kg; it was 460 mm wide, 280 mm high, and 420 mm deep, and its maximum power was 50 W.

The HDTV included both a wide-angle camera and a telephoto camera. For color (RGB) imaging each camera had three separate Panasonic IT charge couple device (CCD) detectors 1.7 cm wide and 2.2 million pixels. Each CCD sensor consisted of $1,920 \times 1,080$ pixels with a pitch of 5×5 μm. The shutter speed of each camera was set either manually or automatically. In automatic mode the shutter speed for each frame was based on the measured brightness distribution of the previous frame.

All images taken by the cameras were digitally compressed (DCT-compressed within the frame) and recorded into two onboard flash memories. Each memory had a capacity of 1 GB, capable of storing 1 min of high-definition video images with the frame rate of 1/30 s in the standard ($1\times$) mode (1,800 frames).

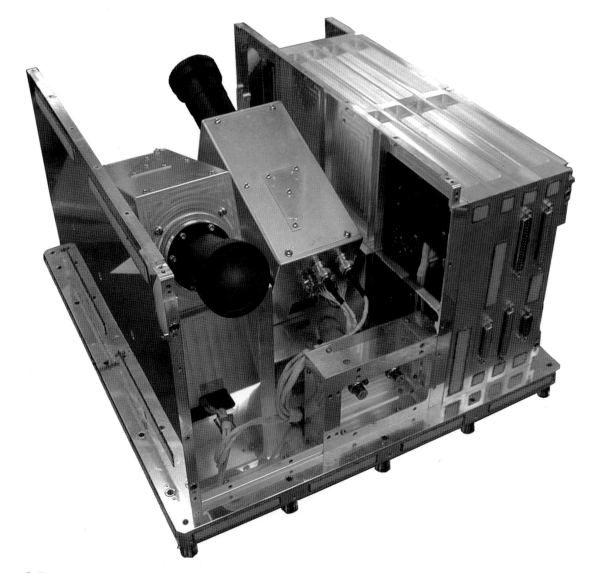

Figure 2.1 Outside appearance of HDTV instrument. The telephoto camera faces this side, and the wide-angle camera faces the opposite side. The box on the right houses the data processing unit (Image courtesy of NHK)

Imaging could also be accomplished in three interval-recording modes. At 2×, 4×, and 8× modes, 2-min, 4-min, and 8-min-long videos were obtained with the frame rate of 2/30, 4/30, and 8/30 s, respectively. With the wide-angle camera the interval mode 8× was generally used for lunar feature observation because coverage of the lunar surface was maximized (ca. 720 km in latitude direction). Apart from video imaging, the HDTV cameras were often used in a still mode. The HDTV cameras can take ten frames at minimum. This mode was useful when download time was limited, such as the period just before *Kaguya's* impact.

The wide-angle and telephoto cameras slanted 22.5° and 18.5°, respectively, below the spacecraft's orbit (Figure 2.2), aiming in the backward and the forward directions of orbital

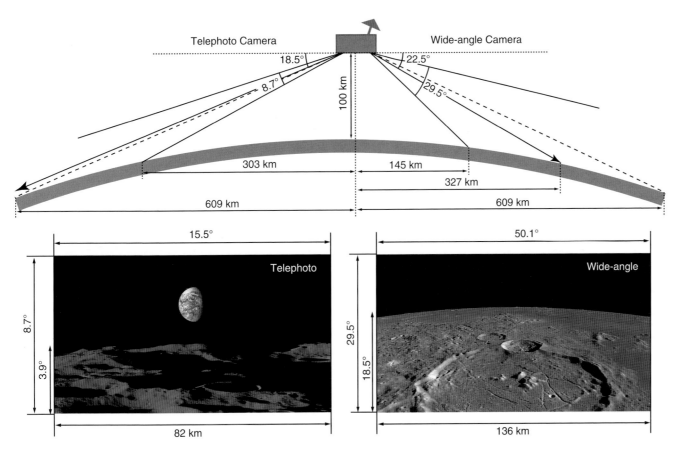

Figure 2.2 Fields of view and areas imaged by the telephoto camera (*left*) and wide-angle camera (*right*) at an altitude of 100 km

motion. The pixel resolutions at the closest point to the Moon (that is, at the bottom of the frame), at the altitude of 100 km, was about 65 and 43 m in the across-track direction and 230 and 658 m in the along-track direction, for the wide-angle camera and telephoto camera, respectively. These are higher spatial resolutions than the Clementine UVVIS and most of the Lunar Orbiter's photographs of the 1960s.

The mission data of the HDTV were transmitted to the ground system in Japan via the X band transmitter on *Kaguya*. Data for a 1-min-long video (about 1 GB) required 20 min to be transmitted to Earth at the rate of about 7.6 Mbps.

Operation

On the way to the Moon, the HDTV telephoto camera captured its first video image of Earth, at the distance of 110,000 km on September 25, 2007 (Figure 3.1). After *Kaguya* entered lunar orbit, the HDTV cameras started imaging in late October 2007.

Kaguya was in a polar orbit of about 2 h period. As it orbited, the Moon rotated 1° in longitude to the west, meaning that the spacecraft returned to the original longitude on the 360th orbit (30th day), and the angle of sunlight on the Moon's surface had changed 30° by that time. Thus, the angle of illumination of topography by sunlight was repeated about 6 and 12 months later. The ascending node of *Kaguya*'s orbit was on the lunar dayside from January to June, and was on the nightside from July to December. *Kaguya* turned over its position by 180° every 6 months, in April and October, to keep its single solar panel pointed at the Sun. For these reasons, the wide-angle camera was aimed northward from April to June and from October to December, and pointed toward the south for the rest of year (Figure 2.3). Consequentially, the HDTV camera needed one full year to image all locations on the Moon under good lighting conditions and with a good pointing direction.

Images of Earthrise and Earthset

True Earthrise and Earthset cannot be seen from most of the lunar surface because Earth is nearly stationary when observed from the Moon. Earthrise and Earthset, in which Earth looks

Figure 2.3 This shows the positional relationship of the Sun, Earth, Moon, and *Kaguya* spacecraft as seen from the north. Beta (β) is the angle between the Sun, Earth, and the orbital plane of *Kaguya*. Omega (Ω) is the angle between Earth, the Moon, and the orbital plane of *Kaguya*. The arrows indicate the pointing direction of the wide-angle camera

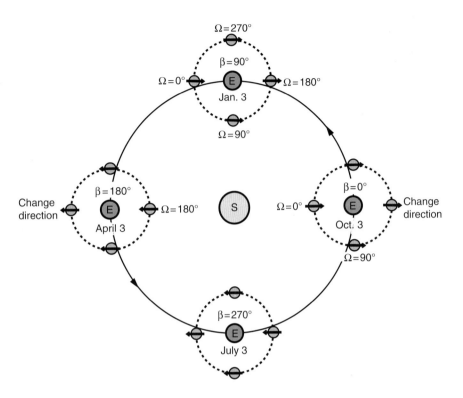

as if it rises and sets below the horizon of the Moon, were seen when *Kaguya* flew over the north and south poles (Figure 2.4), and the spacecraft motion brought Earth into view. The HDTV succeeded in taking Earthrise and Earthset on October 29, 2007, for the first time.

Because the phase of Earth changes continuously as seen from the orbiting Moon there were only two chances for the HDTV to capture a full Earthrise and full Earthset each year, November and April, respectively (Figure 2.3). Full Earth means that nearly 100% of Earth's surface is illuminated by the Sun. The HDTV successfully captured a video of the full Earthrise over the lunar south pole on April 6, 2008 (Figure 3.6).

The HDTV also succeeded in capturing a "Diamond Ring" of Earth from lunar orbit on February 10, 2009 (Figure 3.6). This phenomenon occurs when Earth moves in front of the Sun as seen from the Moon. The Diamond Ring happens only when the outer edge of the Sun is not completely covered by Earth. For *Kaguya*'s image a penumbral lunar eclipse was

Figure 2.4 Configuration of *Kaguya*, Earth, and the Moon in late November 2007 when Earthrise and Earthset images were acquired

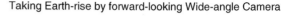

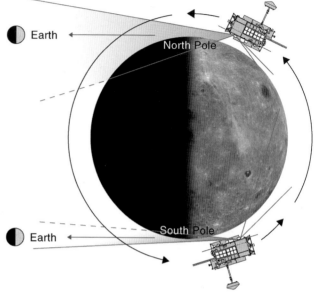

Taking Earth-rise by forward-looking Wide-angle Camera

Earth

North Pole

Earth

South Pole

Taking Earth-set by back-looking Telephoto Camera

seen from Earth. Although *Kaguya* experienced partial and total solar eclipses, the HDTV could not image them because electric power for most of the instruments was available only from the solar panels, which were in darkness during eclipses and could not store the electricity.

Images of the Moon

Because the wide-angle camera and the telephoto camera were slanted toward the bottom of *Kaguya*, the HDTV provided a bird's-eye view of the Moon landscapes, always looking toward the horizon. From an altitude of 100 km the bottom line of the wide-angle camera image covers about 136 km on the Moon (Figure 2.2). This is wide enough to cover most large craters, including Tycho and Copernicus. On the other hand, *Kaguya* also has a powerful "terrain" camera, with a spatial resolution of 10 m and swath width of 35 km at 100 km altitude. The Terrain Camera acquired stereo images that were used to produce a global topographic map of the Moon. Its images, though, are too detailed for most non-scientists. The resolution of the HDTV cameras is between good Earth-based telescopic images and *Kaguya*'s Terrain Camera images and is perfect for giving a true feeling for being in low lunar orbit.

Shadows help to define topography. Most slopes on the lunar surface are less than 30°, so conditions necessary for rich shadow imaging of the equatorial regions occur only in 4 months – January, June, July, and December – when the Sun is low in the lunar sky (Figure 2.3). During the other months, there were few shadows in equatorial areas, but they were plentiful in middle- and high-latitude areas, which were thus targeted for imaging. This resulted from the slant angles of the HDTV cameras so that when the Sun was obliquely behind the spacecraft, landscape features were seen with few shadows, and when sunlight shined obliquely in front of the spacecraft, the landscape was seen with dramatic shadows. Therefore, light conditions of the northern hemisphere and the southern hemisphere of the same latitudes are different (Figure 2.5a–c). The relative qualities (i.e., "good," "moderate," or "bad") of the imaging conditions for different latitudes in both cases described above are listed in Table 2.1. We tried to select the optimum period to image each selected feature.

The number of HDTV images acquired each month is shown in Figure 2.6. During the nominal operation period of the scientific instruments (from December 12, 2007, to October. 31, 2008), fewer than ten images per month could be taken because the HDTV camera's priority was the lowest among all the mission instruments. But before and after that period, much imaging was able to be performed. Altogether, 616 videos and 21 still images of the Moon were taken; their coverage is shown in Figure 2.7 (below).

Still Images from HDTV

HDTV videos are spellbinding but do not translate easily into still images. Two methods were employed to create images for this book. Dr. Rie Honda of Kochi University introduced the first method. A video sequence of HDTV contains 1,800 frames, each 1,080 pixels high by 1,920 pixels wide. Dr. Honda selects one specific horizontal line 1,920 pixels

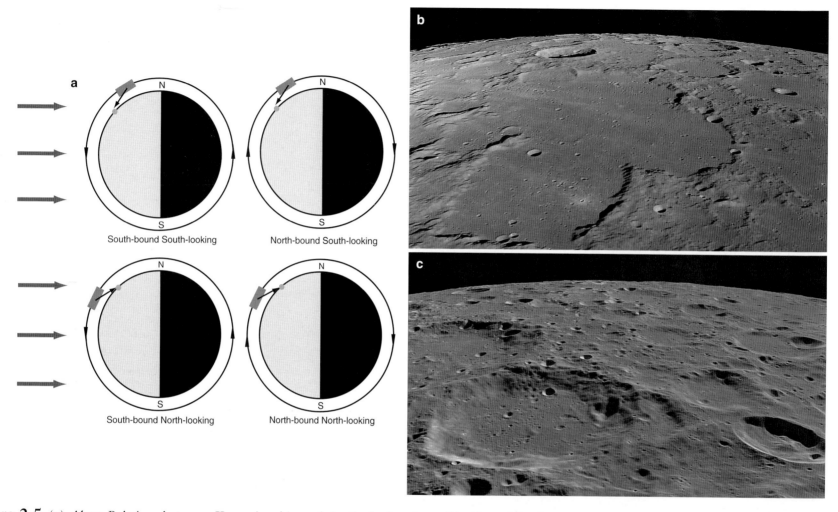

Figure 2.5 (**a**) *Above*: Relations between *Kaguya*'s orbits and the Sun's elevations. (**b**) *Above right*: Crater Meton (70°N), Sun elevation 20°; wide-angle camera facing the north pole. (**c**) *Below right*: Crater Neumayer (70°S), Sun elevation 20°; wide-angle camera facing the equator

Table 2.1 Light condition of HDTV wide-angle camera

	Jan.	Feb.	Mar.	Apr.	May	June	July	Aug.	Sep.	Oct.	Nov.	Dec.
60°N~90°N	M	G	M	M	M	M	M	G	M	M	M	M
30°N~60°N	G	G	–	–	M	G	G	G	–	–	M	G
30°S~30°N	G	M	–	–	M	G	G	M	–	–	M	G
60°S~30°S	G	M	–	–	G	G	G	M	–	–	G	G
90°S~60°S	M	M	M	M	G	M	M	M	M	M	G	M
Local time	Morning		Daytime		Evening		Morning		Daytime		Evening	

G: good; M: moderate; –: bad

wide for every image. When these specific lines from 1,800 images are placed one above the other, a rectangular image of 1,920 pixels by 1,800 pixels results. If this image is stretched in the appropriate ratio, an image such as Figure 2.8a results. Such images are excellent to show the coverage of *Kaguya*'s HDTV and as an index to *Kaguya* HDTV videos.

The second method, conceived by one of the authors (Shirao), is simpler than the first one. The bottom part of an HDTV image continuously disappears as *Kaguya* moves above the lunar surface. To create a still image Shirao chose one image first, then connected it to the lower part of the previous image (Figure 2.8b). Then he would magnify the connecting image to match the scale of the bottom part of the first image. The magnification ratio is different horizontally and vertically, because the look angle changes gradually when approaching the topography. This action was repeated several times using Photoshop CS4, producing the "wide-view" images as in Figure 2.8c. This piecing together of strips from subsequent video frames accounts for the stair-step edges of the images in this atlas.

Strictly speaking our "wide-view" still images have several faults. Their look angle is a little different from the true one. Although magnification of each segment is constant on wide-view images, magnification of real wide-angle images changes gradually. Additionally, some pixels overlap or may be missing at the junctions in the wide-view images. Despite all of these defects, our wide-view images provide spectacular astronaut-eye panoramas of 100-km-wide swaths of lunar topography all the way to the horizon.

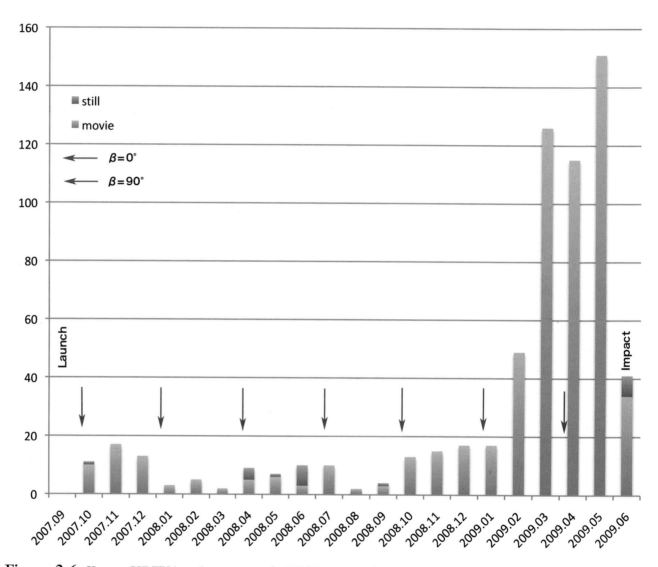

Figure 2.6 *Kaguya* HDTV imaging per month. HDTV acquired most of its imaging after the nominal mission was completed in 2009

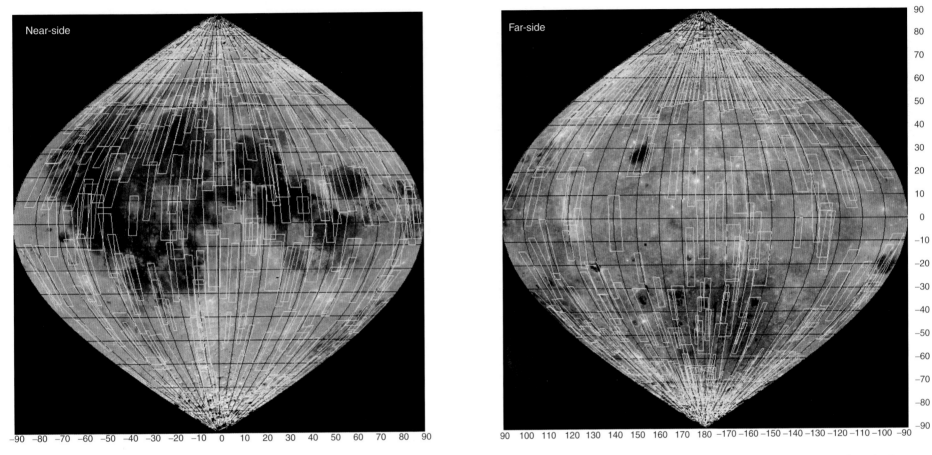

Figure 2.7 Yellow rectangles depict coverage of movies by wide-angle camera; blue boxes mark still images by wide-angle; and purple ones are movies by telephoto camera (Image courtesy of R. Honda)

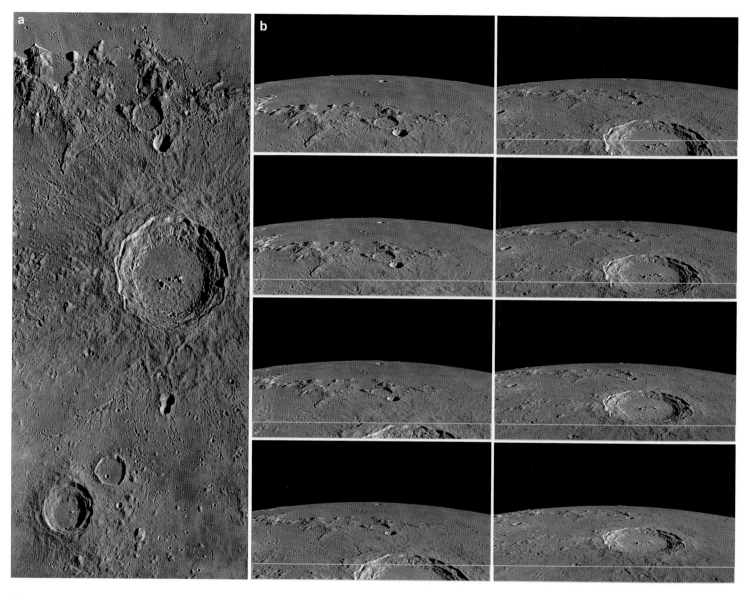

Figure 2.8 (**a**) An example of strip images created by Honda's method. (**b**) a sequence of trimmed images to create a wide-view image

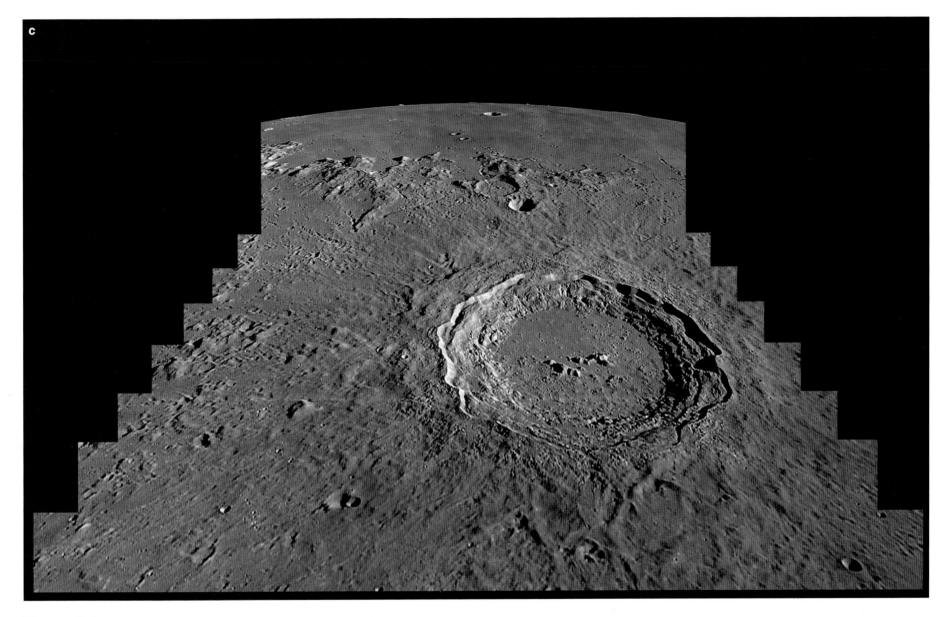

Figure 2.8 (c) finished wide-view image by Shirao's method

Images of Earth and *Kaguya*'s Impact on the Moon

CHAPTER

3

Figure 3.1 Earth from 110,000 km

The HDTV system started taking images on September 29, 2007, and this is one of the first. At 110,000 km from Earth, the telephoto camera took an image of the "receding Earth" in from a transfer orbit with *Kaguya* heading to the Moon. The western edge of South America is clearly shown.

Figure 3.2 Earthrise as seen by the wide-angle camera

This is Earthrise over the lunar south pole as recorded by the wide-angle camera. The bottom of the image shows the Moon at about 85°S latitude. The 19-km-wide distinct crater in the middle is Shackleton, along whose rim is the lunar south pole. The ridge on the terminator is Malapert's south rim, the other side of which we can observe with telescopes from Earth (October 1, 2008).

Figure 3.3 Earthset over the lunar south pole

From the Moon, Earth goes through phases, just as the Moon does when viewed from Earth. This is the 9.8-day-old Earth. Earthrises and Earthsets captured by *Kaguya* were always seen near the lunar north or south poles. This is why Shackleton is visible so often in the foreground. In this image of Earth, the Atlantic Ocean is in the center, Antarctica is on the top, and the Sahara Desert is in the lower left corner (December 4, 2007).

Figure 3.4 Thin crescent Earth

This is the very thin crescent Earth, 9 h 8 min before the "New Earth", as seen from over the lunar north pole. The arc of the illuminated Earth exceeded more than 180° owing the scattering of sunlight by Earth's atmosphere (March 22, 2008).

Figure 3.5 This is a composite of telephoto camera images of Earth at 2-h intervals over the lunar south pole from January 27, 13:32 (UT) to January 28, 11:04 (UT) in 2008

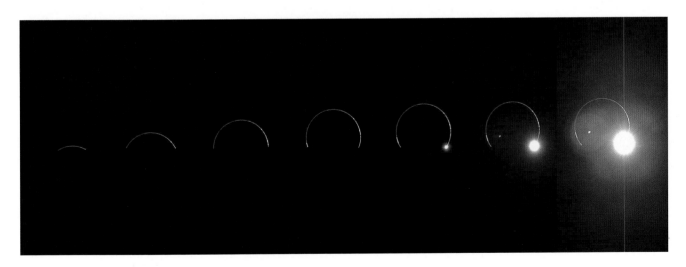

Figure 3.6 Diamond Ring of Earth, imaged on February 9, 2009, when a penumbral lunar eclipse viewed from Earth appeared as a partial solar eclipse when seen from the Moon

Images of Earth and Kaguya's Impact on the Moon

27

Figure 3.7 One day of the gibbous Earth

This composite of telephoto camera images captures Earth at 2-h intervals over the lunar south pole from May 2, 20:21, to May 3, 20:09, 2008. Because the orbital period of *Kaguya* is about 2 h, the surface of Earth rotated from left to right. You can identify Australia, India, Saudi Arabia, Africa, and North and South America. South is up.

Images of Earth and
Kaguya's Impact
on the Moon
29

Figure 3.8 Full Earthrise

Kaguya's HDTV telephoto camera captured a 1-min movie of the "full Earthrise" on April 7, 2008, when *Kaguya* was flying over the lunar south pole. This image showing Earthrise is a composite at 10-s intervals of the movie from the left to the right. In the image of Earth, the Pacific Ocean is visible in the center, with part of North America at the lower left.

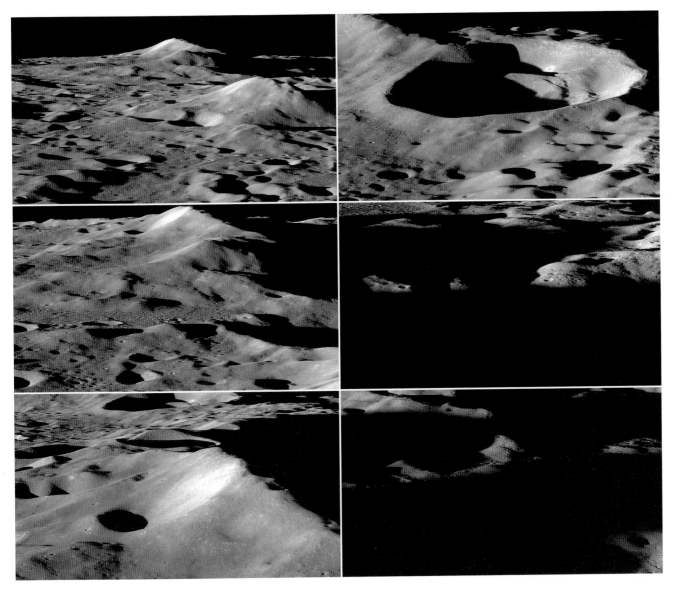

Figure 3.9 (a) *Kaguya* impact of the Moon

Kaguya impacted on the near side at 80°E and 64°S near the crater Gill, on the dark side close to the southeastern limb. These consecutive still images were obtained just before the controlled impact on the Moon at 18:25 on June 10, 2009 (UT). The forward-looking telephoto camera took the images at 1-min intervals from 18:11 to 18:16; so, we can see the approaching lunar surface. The last image was pointed 81°S and 261°E from 20.7 km high.

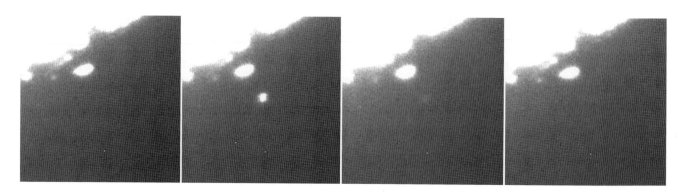

Figure 3.9 (b) *Kaguya* impact as seen from Earth

The impact of *Kaguya* on the Moon was observed with the Anglo-Australian 3.9-m telescope and an infrared camera at Siding Spring, Australia. A series of 1-s exposures was taken with 0.6 s between exposures. This image shows a sequence of four frames around the impact time, with a bright impact flash visible in the second frame and faintly seen in the third and fourth. The large bright blob that does not change is a mountain peak illuminated by the Sun. (Image courtesy of the University of New South Wales/Anglo-Australian Observatory; J. Bailey and S. Lee.)

The Lunar Surface

The Moon is the most remarkable object in the nighttime sky. It is, of course, much smaller and less impressive in absolute dimensions than planets such as Mars and Jupiter and is hardly worthy of mentioning in the same sentence as a galaxy. But the Moon is more visibly interesting to a person on Earth with eyes, binoculars, or telescope than any other object in the universe. The interest increases as you get closer to the Moon, and in the 1960s and early 1970s *Apollo* astronauts in low lunar orbit captured many evocative photographs. These differed from most other orbital images of the Moon that were taken by robotic spacecraft, looking straight down. Such vertical images provide a very utilitarian map view, but lack the sense of three-dimensionality of the astronaut's perspective. Now, using the first high-definition TV camera flown on a lunar spacecraft, images from the Japanese *Kaguya* lunar orbiter provided a perspective identical to what the astronauts experienced, but covering far more area than the *Apollo* astronauts saw. With this atlas we share this perspective with everyone.

Although the images themselves are often spellbinding, appreciation increases with understanding of what is depicted. To help learn how to read the pictures a short essay accompanies each image. The goal is not to identify every landform, but to discuss a few as examples of general processes that formed and modified the Moon. As you read the text and look closely at each image you will be able to interpret more in the images than is described. You will be learning to read the Moon.

Impact Craters and Basins

The Moon is a relatively simple world with only two geologic processes important in its development. The dark patches visible during full Moon are giant piles of lava flows that erupted mostly between 3.6 and 2.5 billion years ago. The bright areas of the Moon are

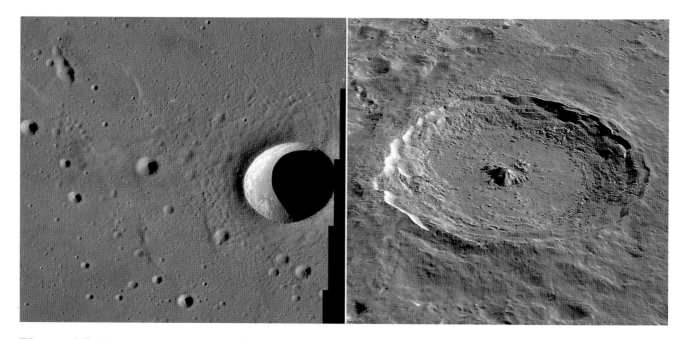

Figure 4.1 Simple crater Rümker E, diameter 7 km (*Kaguya* HDTV image)

Figure 4.2 Complex crater Tycho, diameter 83 km (*Kaguya* HDTV image)

older surfaces that are intensely pitted by impact craters. When comets, asteroids, or other debris from the formation of the Solar System smashed into the Moon, with velocities of kilometers per second, holes were excavated and ejecta was thrown out in all directions.

The shapes of craters depend on how big they are. Impact craters with diameters smaller than about 15 km are simple bowls with steeply slanting walls that lead to small flat floors (Figure 4.1).

Impacts of larger projectiles result in a more complex crater morphology. During the impact process the walls of craters larger than about 15 km diameter begin to fail, with pieces landsliding down slope to the crater floor. The larger the crater, the more massive the piles of material that slump off the walls. These individual collapses often take a bite out of the rim, leaving a scalloped edge with mounds of debris below.

At some crater diameter – typically by about 40 km – entire sections of the rim give way and slide down as coherent annular blocks or terraces. Complex craters such as Tycho (Figure 4.2)

Figure 4.3 The younger crater Aristoteles (diameter 87 km) sliced through the rim of the older and smaller Mitchell (diameter 30 km, on the left) (*Kaguya* HDTV image)

and Copernicus are wreathed with massive terraces that stair-step down from the rim crest to the floor. Floors for these larger craters are also different than those of simple craters. For larger craters, floors are wider and central peaks appear, caused by a rebound of compressed rocks at ground zero of the impact. For smaller complex craters, there is often a single peak, and for larger craters massive clumps of mountains rise above the floor.

The pristine appearance of craters changes over time, sometimes instantaneously, but most often slowly. A crater that happens to be where a more recent, larger one forms will be totally destroyed. A new crater may cut through an older one's rim or partially cover it with ejecta (Figure 4.3). Moonquakes caused by impacts shake craters, causing rocky material to move down slope, smoothing out terraces until just a bumpy wall remains. Infill by ejecta and down slope avalanching of rocks turn deep craters into shallow ones.

Craters along the edges of the maria often have lava-flooded floors but intact walls, implying that the lava rose up through fractures beneath the crater (Figure 4.4). Other craters have breached walls, and giant lava-falls must have cascaded onto their floors, covering the central peaks and sometimes filling the entire crater. In some cases only the top of the rim remains, and for other craters this is buried, too, leaving a low circular ridge in the mare to mark a buried crater. Look closely near the terminator – the day-night shadow zone – to detect such ghost craters.

Figure 4.4 Goclenius (*upper right*, 72 km diameter) and Magelhaens (*bottom left*, 40 km) have lava-flooded floors (*Kaguya* HDTV image)

The most interesting craters on the Moon occur near the edges of the maria and have been invaded by lava. In these cases the crater rims have not been breached, and lava rose up the fractures created by the impact that formed the craters. Often the magma formed pools in these fracture zones, exerting a pressure that bodily raised the crater's floor. Usually, concentric fractures and rilles formed near the edges of the floor, and sometimes small amounts of lava leaked to the surface. Gassendi, Posidonius, and Atlas (Figure 4.5) are prominent examples of floor-fractured craters.

Very large impact events create craters hundreds or a thousand kilometers in diameter. The interiors and exteriors of these landforms are so different from normal craters that they have a different name – basins. There are more than 50 basins on the Moon, and these are among the oldest lunar features and have been strongly modified so that sometimes they are difficult to detect. Although they were originally deep depressions, many are filled with mare lavas. In fact, the maria have sometimes been called circular maria because most occupy impact basins and thus have a circular shape. Mare Crisium's circular outline is visible even with a naked eye, and closer looks with binoculars or small telescopes reveal the mountainous basin rim that surrounds it. The circularity of maria like Serenitatis, Tranquillitatis, Imbrium, and Humorum attest to their underlying basins.

Figure 4.5 Floor-fractured crater Atlas (diameter 50 km) with rilles and a dark pyroclastic deposit on its floor (*Kaguya* HDTV image)

Just as the progression from simple craters to complex craters is marked by the development of terraces and central peaks, these features are strongly transformed in basins. The first morphological feature to appear in small basins is a mountain ring on the basin floor, between the central peak and the main wall. In larger basin diameters the central peak disappears and multiple rings form, both inside and outside the most prominent rim. These features are best displayed in the Orientale basin (Figure 4.6). Because this basin is on the very edge – or limb – of the Moon as seen from Earth it was difficult to understand its morphology. But the US Lunar Orbiter IV spacecraft obtained a stunning image of the basin in 1967, and the structure of the basin was revealed.

Orientale is a multi-ring basin whose concentric mountains make it look like a bull's-eye target. The most conspicuous ring is named the Cordillera Mountains, with a basin diameter of about 930 km. Faint rings have been mapped outside the Cordilleras, but the three inner rings are much more visible. Two circular mountain chains make up the Rook Mountain rings, and an inner ring is defined by the edge of the small mare. Inner rings such as the Rook are visible in other basins, but because most are filled with mare lavas often only the tallest peaks are seen.

The exterior of Orientale is as spectacular as the inside. The natural excavation of a giant hole that is now 7–8 km deep required the removal of billions of tons of rocks. That material

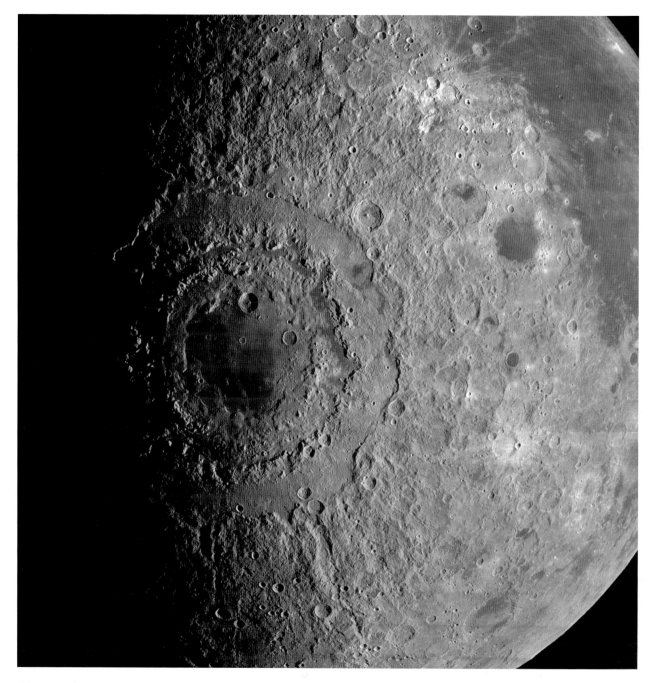

Figure 4.6 Orientale impact basin, 930 km diameter (NASA Lunar Orbiter IV image)

Figure 4.7 Secondary crater chains on the lunar farside, created by ejecta from Orientale Basin (*Kaguya* HDTV image)

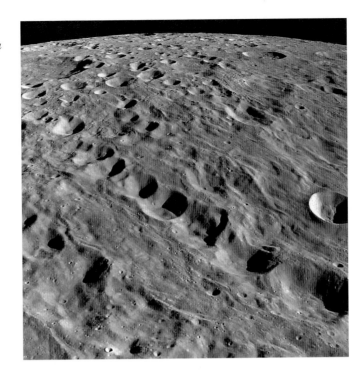

was fractured, pulverized, melted, and ejected beyond the basin's rim. Thick deposits of ejecta outside the Cordillera completely obliterated most nearby preexisting craters, covering them with pasty material that flowed across the surface. Much of the more distant ejecta appears to have been so heavily pulverized that when it settled on hill slopes it flowed downward, filling in crater floors and other low spots with smooth, relatively bright material. Further away small mountains that were ejected along ballistic paths created lines of secondary craters 10–20 km in diameter (Figure 4.7). When Orientale formed, about 3.8 billion years ago, some ejecta probably fell on the Moon and some escaped lunar gravity and pelted Earth with debris, forming many significant craters on Earth. It is a good thing that this was long ago, for it would have been deadly to much of Earth's early life.

Lunar Volcanism

The dark maria are the largest volcanic landforms on the Moon. Nearly all of these mare lavas erupted on the hemisphere facing Earth, and most are in the northern hemisphere within and overflowing a half-dozen basins (Figure 4.8). Millions of volcanic eruptions

Figure 4.8 Mare Serenitatis lavas fill an underlying ancient basin about 920 km wide, the same size as the Orientale Basin but much older (Image by Motomaro Shirao)

occurred from about 3.6 to 2.5 billion years ago, creating the great piles of lava we see today – the maria. Small amounts of volcanic activity occurred earlier, and some lavas flowed across the surface as recently as 1 billion years ago. Volcanism stopped because the Moon's interior cooled enough for the mantle and at least part of its core to solidify.

Although the lava flows themselves are almost featureless, many smaller landforms associated with them are among the most interesting structures on the Moon. This includes sinuous rilles, domes, collapse pits, and dark halo craters. Lava moves in flows, and the center of each flow moves fastest, often creating a channel within levied walls. Such channels are commonly seen on Earth's volcanoes, so the similar channels – or sinuous rilles – on the Moon are well understood. *Apollo 15* landed near the Hadley Rille (Figure 4.9), one of the Moon's most famous sinuous rilles, to explore its structure. In many lava flows both on Earth and the Moon a channel's surface solidifies while lava keeps flowing beneath it. When the lava finally drains, a hollow tube is left behind. Sometimes pieces of a tube's roof fall in, creating collapse pits. On Earth, cave explorers use these skylights to enter lava tubes, and someday the same thing will happen on the Moon.

Figure 4.9 Hadley Rille, a sinuous lava channel (*Kaguya* HDTV image)

Some eruptions produce low mounds of lava called domes. These also are familiar from Earth's volcanic regions, and result from an eruption on flat ground so that flowing lava shifts from one low spot to the next, gradually spreading out in all directions and building a rounded hill. Typically a dome has a volcanic crater at its summit, marking the vent where lava emerged onto the lunar surface (Figure 4.10)

Nearly all of the Moon's volcanic landforms were made by smoothly flowing lava, but in some cases the rising magma contained enough gas to make bubbles that burst apart the magma, erupting pyroclastic or ashy material. Almost all of the volcanic dark halo craters – called that because the ash makes a dark ring around their vents – are found on the floors of floor-fractured craters (Figure 4.11). In some cases ash also explosively erupted from rilles, creating dark patches of pyroclastics up to a few hundred kilometers across.

Faults

The crust of Earth is cut by many faults, both large and small, associated with the movement of the geologic plates of crust and upper mantle rocks. Such plates never formed on the

Figure 4.10 A small dome about 15 km in diameter in northern Oceanus Procellarum with elongated summit collapse pits (*Kaguya* HDTV image)

Moon, and thus our satellite has far fewer faults. But there are some, which fall into three classes. The inner wall of every crater and basin has had rocks move downward toward the floor. In many cases a section of the material broke off the wall and slid down as a giant terrace. The collapsed walls or scarps of basins are the largest faults on the Moon. Surrounding a portion of the Nectaris Basin is the Altai Scarp edge of that basin's rim, and the front of the Apennine Mountains (Figure 4.12) is a similar fault scarp of the Imbrium Basin rim.

Only a few other faults have been found on the Moon, most famously the Straight Wall (Figure 4.13), near the eastern edge of the Nubium Basin. The Wall occurs in the middle of a crater that once stood on Nubium's rim. As the center of the basin collapsed the crater's seaward wall was carried down and ultimately covered by lava, and the crater's floor fractured. The eastern side of the fault is the high side, representing the crater's original floor, and west of the fault the surface is about a half kilometer lower and covered with subsequent lavas.

There is one other class of fault that often is unrecognized. The circular maria have circumferential ridges. The origin of these was debated – perhaps they were volcanic or maybe folding of the surface – until *Apollo* images and data were used to explain them as faults.

Figure 4.11 Three dark patches on the floor of the crater Alphonsus are pyroclastic deposits, ashes from the small pits that erupted at their centers (*Kaguya* HDTV image)

Figure 4.12 The Apennine Mountains are part of the rim of the Imbrium impact basin and the bright scarps mark faults where rocks slid onto the basin floor, later to be covered by lava flows (*Kaguya* HDTV image)

Consider a basin to be a bowl of solidified lava. If the basin subsides – because of the weight of the lava on its floor – the lava is pulled down into a smaller area of the basin. The solid lava pile has to deform to fit into the smaller available volume, and so it fractures, with sections of the lava sliding up over adjacent pieces. These low angle faults create mare ridges (Figure 4.14).

Figure 4.13 The Straight Wall is 110 km long, and its high right side rises about 500 m above the left side (*Kaguya* HDTV image)

Figure 4.14 Mare ridges in northern Oceanus Procellarum near Rümker (*Kaguya* HDTV image)

Part II
The Atlas

ORGANIZATION OF THE ATLAS

This atlas illustrates 100 of the Moon's most interesting landforms. It does not cover the entire lunar landscape but highlights features that are in many cases the best the Moon has to offer. Atlas pages are arranged geographically – or more correctly, selenographically – starting with the crater Gauss in the northeastern region of the side of the Moon visible from Earth. Subsequent images follow the limb southward to the crater Janssen in the southeast. Then, like a person mowing their grass, the progression of images turns around and heads north, picking up interesting features just to the west of the previous swath. On reaching the north pole, another 180° turn follows, and the next swath moves southward near the lunar prime meridian. This back and forth progression across the surface continues to the western limb, near Orientale, and then presses on across the far side. Thus, pages before and following any given page are likely to depict features relatively nearby, but overlap is rare. Seventy-seven images show objects on the nearside of the Moon, and the remainder depict farside features. Scientists and amateur astronomers are lucky that the lunar maria and all their fascinating landforms are on the side of the Moon that faces us.

Each atlas page shows a portion of a *Kaguya* HDTV pass, oriented with the horizon at the top. This means that sometimes north is up and other times south is. This does not matter for viewing and interpreting the scene but is important for comparison with maps and other images. The small lunar image on each page is a location map with an arrow indicting the position and look direction of the HDTV image. For example, the arrow on atlas sheet 1 points up so that page's image has north at the top. On sheet 6 the arrow points down, indicating that the horizon at the top of that sheet is toward the south. The arrows also point to the center of each image, providing a location context. Note that the location maps – made from the *Clementine* global mosaic – are centered in six orientations: Earth side, far side, northern hemisphere, southern hemisphere, eastern limb, and western limb.

The scale of the images is not constant because the altitude of the spacecraft above the surface varied from 21 to 116 km. Of course, since the images depict the surface all the way to a horizon, the scale changes from near to far. Additionally, two of the images (#30 and #99) were obtained with the telephoto lens and have a different scale and more limited field of view. Generally, the size of a feature is identified in its caption to provide scale.

The location of each atlas plate is indicated on the following six maps by numbers. The maps are not photographs but were created from more than 6 million digital elevation data points collected by the *Kaguya* altimeter. Using this data it is possible to depict every place on the lunar surface with a consistent illumination angle so that all parts are equally well depicted. The first four maps are centered on the eastern limb (270°), the nearside (0° longitude), the western limb (90°), and the far side (180°). These topographic depictions are merged with surface brightness or reflectivity data from the *Clementine* spacecraft to create a natural appearance. Similar brightness images are not available from *Clementine* for the lunar poles, so the last two images use color to depict elevation, with yellows and red being high terrain and blue and purple low.

Details of the location of each image, and date and other circumstances of its acquisition, are provided in the Thumbnail Index near the end of the book.

N

Hayn

Endimion
MARE
HUMBOLDTIANUM

Compton

Fabry

1 Gauss   100 Giordano Bruno

Cleomedes 2

90

MARE
MOSCOVIENSE

Joliot

MARE
CRISIUM

4 3

MARE
MARGINIS

5

Fleming

MARE
TRANQUILLITATIS

Neper

98

King

Mendeleev

7

MARE
FECUNDITATIS

6
MARE
SMYTHII

97

8

Langrenus

18

20

Pasteur

19

9

Hilbert

99

17

Petavius

10

11
Humboldt

96

Scaliger

Tsiolkovskiy

Gagarin

13
Stevinus

Furnerius

12

93

14

Vallis Rheita

MARE
AUSTRALE

16

95

Planck

15   94 Schrödinger

93

S

Figure Eastern Hemisphere. In all six maps elevation data are from the Kaguya LALT team, NAOJ/ JAXA, and the albedo data of the first four are from Clementine/NASA

Near-side

Far-side

N

30
28
31
MARE FRIGORIS
34
55
32
Aristoteles
56
Plato
33
27
Atlas
Sinus Iridum
57
54
53
Posidonius
Cleomedes
Archimedes
26
MARE IMBRIUM
59
35
2
MARE
SERENITATIS
25
Aristarchus
60
Montes
Apennius
37
MARE
CRISIUM
36
4
OCEANUS PROCELLARUM
MARE
VAPORUM
3
61
50
Eratosthenes
23
Marius
48
24
Copernicus
38
Kepler
52
7
62
49
39
40
51
MARE
TRANQUILLITATIS
Sinus Medii
22
W
E
MARE
FECUNDITATIS
47
8
9
21
18
63
41
20
Ptolemaeus
Theophilus
Alphonsus
19
Gassendi
MARE
NECTARIS
64
Bullialdus
46
MARE
NUBIUM
MARE
65
17
HUMORUM
68 66
42
10
65
43
13
67
12
44·45
14
69
Tycho
16
Schickard
70 71
Clavius

S

Figure Nearside

Birkhoff

Pythagoras

55

56

57
Rümker

58

Landau

Kovalevskaya
Lorentz

59

60
Aristarchus

OCEANUS PROCELLARUM

Poynting

61 Marius

Vasco Da Gama

Michelson

62

80

Copernicus

52

Hertzsprung

Hevelius

Riccioli

Korolev

Grimaldi

63

MARE
OLIENTALE

77

75

79

64
MARE
HUMORUM

65

76

Montes Cordillera

66

68

67

82

Chebyshev

74

70

69 Schickard

Apollo

Mendel

78
Wargentin

Hausen

73

Bailly

N

S

Far-side

Near-side

Figure Western Hemisphere

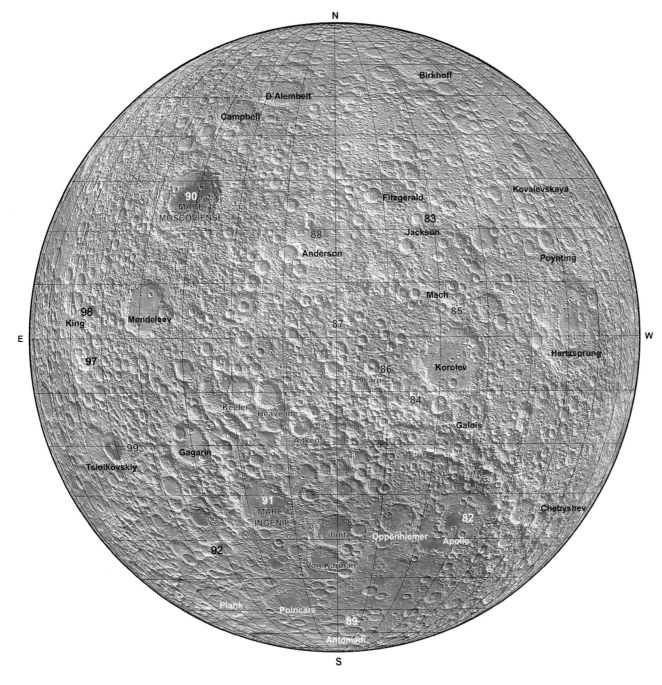

Figure Farside

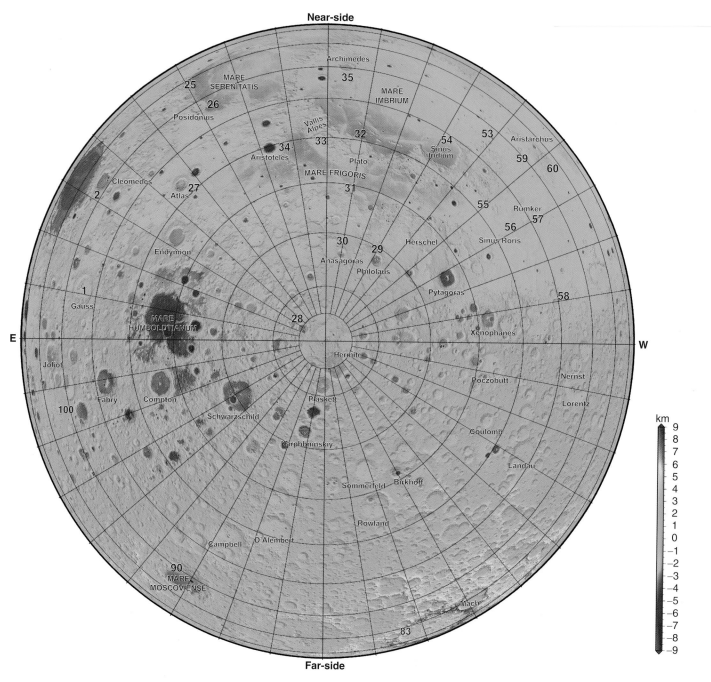

Near-side

Archimedes
35
MARE
SERENITATIS
25
26
MARE
IMBRIUM
Posidonius
Vallis
Alpes
32
54
53
Aristarchus
34
33
Sinus
Iridium
59
Aristoteles
Plato
60
MARE FRIGORIS
Cleomedes
27
31
2
Atlas
55
Rümker
56
57
30
Herschel
Sinus Roris
Endymion
29
Anasagoras
Philolaus
1
Pytagoras
58
Gauss
28
MARE
HUMBOLDTIANUM
Xenophanes

E   W

Hermite
Joliot
Poczobutt
Nernst
Compton
Plaskett
Coulomb
Lorentz
Fabry
100
Schwarzschild
Landau
Karpbhninskiy
Sommerfeld
Birkhoff
Rowland
Campbell
D'Alembert
90
MARE
MOSCOVIENSE
Mach
83

Far-side

km
9
8
7
6
5
4
3
2
1
0
−1
−2
−3
−4
−5
−6
−7
−8
−9

Figure North Pole

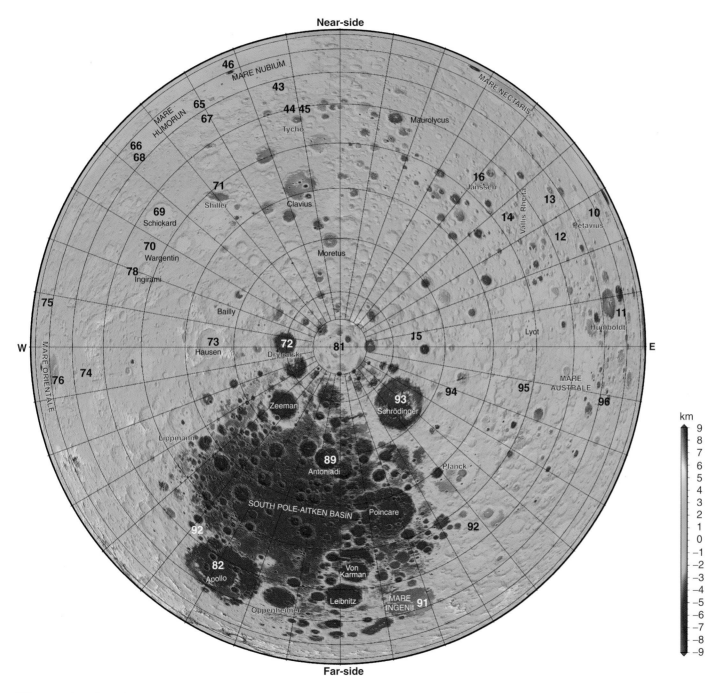

Figure South Pole

Lunar Names

To talk meaningfully about a place it is necessary to have names for important landforms. The present system of lunar nomenclature is descended from the Jesuit astronomer Riccioli, who in 1651 added names to features depicted on a lunar map made by his student Grimaldi. Riccioli used Latin names for the maria that reflected the idea that the Moon influenced Earth's weather. Thus, there are maria Frigoris (sea of cold), Humorum (moisture), Imbrium (showers), and Nubium (clouds). A few others were named for states of mind – Serenitatis (serenity) and Tranquillitatis (tranquility) – and or other things (Nectaris from nectar and Crisium from crises).

Riccioli named craters after famous astronomers, with the ancient Greeks in the north and more recent scientists toward the south. Not being modest, he placed his name and Grimaldi's on large craters near the western limb. Since Riccioli's time many more names have been added to the Moon so that now there are about 1,240 on the side facing Earth and another 690 on the far side. The International Astronomical Union has assigned names to lunar features for more than 80 years, with the requirement that the person commemorated was a prominent scientist who has been dead at least 3 years. Scientists may hope to receive the honor of having a crater named for them, but not too soon!

There are many more craters on the Moon than deserve a name, so many just have letter designations. Thus, craters near Copernicus might be Copernicus A, Copernicus B, etc.

Since many lunar books in the nineteenth century were in English, the official names for features other than craters were in English. This changed in 1960, when Latin was introduced for lunar terms. Because both the English and Latin names are still widely used, we include both in the index but use the English names in the text. Here are translations of the most common Latin terms for types of lunar landforms mentioned in this book.

Catena – crater chain
Dorsum – mare ridge
Lacus – lake
Mare – sea (pl. maria or seas)
Mons – mountain, (pl. montes or mountains)
Oceanus – ocean
Palus – marsh
Promontorium – cape
Rima – rille
Rupes – scarp
Sinus – bay
Vallis – valley

Plates 1 to 28

1: Gauss

Virtually all lunar craters were formed by impact, and although many look similar, each is unique. The interior of the crater Gauss offers examples and explanations. The 177-km-diameter Gauss is a floor-fractured crater – a class that includes many of the most fascinating craters on the Moon. It is old enough that its original wall terraces have been smoothed, its floor has been flooded by lavas (hiding most of its central mountains), and the lava has been cracked with circumferential rilles, one of which erupted ash, creating a dark halo crater. The three largest and somewhat dingy craters on the southern floor formed before the lava flooded the big crater. These craters were surrounded by lava, which entered where a wall is missing. A bright and crisp-rimmed simple crater interrupts a rille and thus is younger than the rille and the lava flooring it helped form.

2: Cleomedes

The north shore of Mare Crisium (bottom of image) is fringed by basin ejecta that has been reworked by the formation of many impact craters. The largest is Cleomedes, a 125-km-diameter crater that looks like a miniature mare itself. Its flat floor is covered with lava that buried all but an off-center ridge of the original central mountains. Four medium-size impact craters cut into the floor, but the most interesting features are the thin rilles that make straight creases across the right side of the floor. Just visible near the rilles are small craters surrounded by dark halos. These dark halo craters are small volcanoes – like cinder cones on Earth – that erupted dark ash. The rilles mark subsurface vertical sheets of magma that probably fed the lava flows that cover the floor and occasionally erupted as explosions of ash.

Looking across the western edge of Mare Crisium there is a strong contrast between the relative smoothness of the mare and the surrounding highlands. The highlands include massive blocks of ejecta from the formation of the huge impact basin that later lavas filled. The low and narrow mare ridge that curves from the bottom right concentric t o the shore is the boundary between thick lava fill to the right and a shallow lava-covered bench to the left. Lick, the crater partially seen at bottom right, Yerkes (above it), and the curved ridge remnant of another crater north of Yerkes are all craters that formed on an earlier floor of the Crisium Basin. Later mare lavas surrounded and flooded these craters but did not completely cover them because they formed on an annular bench that is much shallower than the deeper center of the basin. Similar craters must have formed in the basin's interior, but these were completely covered by mare lavas.

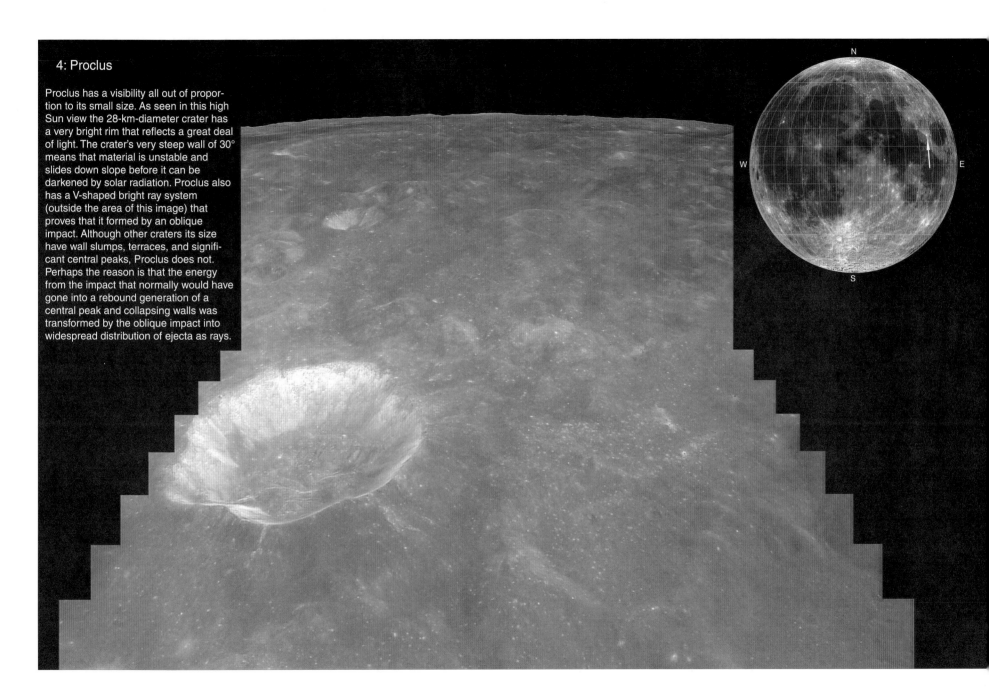

4: Proclus

Proclus has a visibility all out of proportion to its small size. As seen in this high Sun view the 28-km-diameter crater has a very bright rim that reflects a great deal of light. The crater's very steep wall of 30° means that material is unstable and slides down slope before it can be darkened by solar radiation. Proclus also has a V-shaped bright ray system (outside the area of this image) that proves that it formed by an oblique impact. Although other craters its size have wall slumps, terraces, and significant central peaks, Proclus does not. Perhaps the reason is that the energy from the impact that normally would have gone into a rebound generation of a central peak and collapsing walls was transformed by the oblique impact into widespread distribution of ejecta as rays.

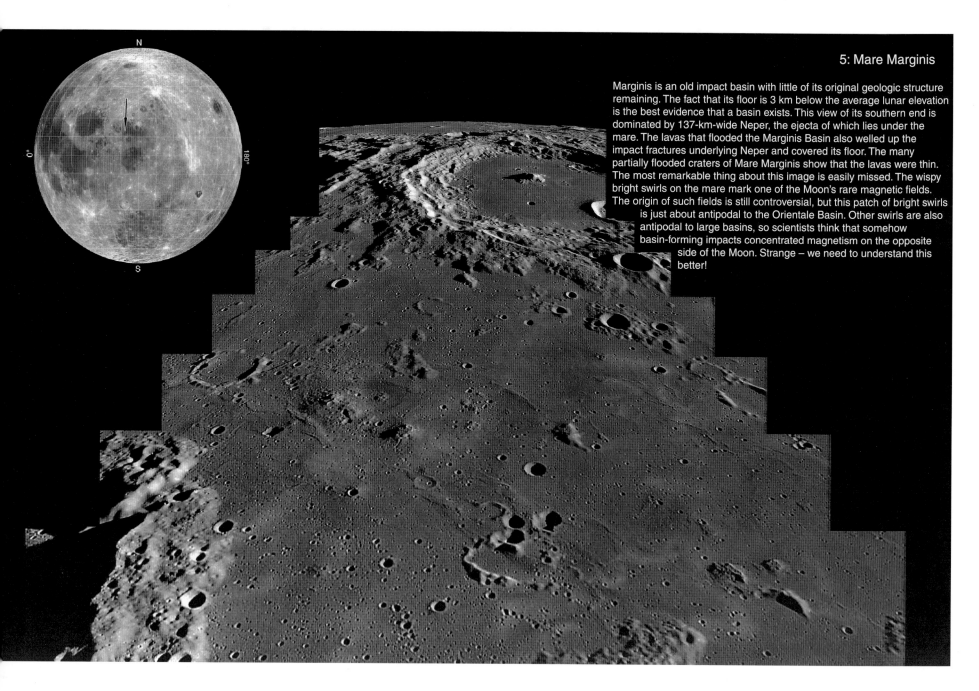

5: Mare Marginis

Marginis is an old impact basin with little of its original geologic structure remaining. The fact that its floor is 3 km below the average lunar elevation is the best evidence that a basin exists. This view of its southern end is dominated by 137-km-wide Neper, the ejecta of which lies under the mare. The lavas that flooded the Marginis Basin also welled up the impact fractures underlying Neper and covered its floor. The many partially flooded craters of Mare Marginis show that the lavas were thin. The most remarkable thing about this image is easily missed. The wispy bright swirls on the mare mark one of the Moon's rare magnetic fields. The origin of such fields is still controversial, but this patch of bright swirls is just about antipodal to the Orientale Basin. Other swirls are also antipodal to large basins, so scientists think that somehow basin-forming impacts concentrated magnetism on the opposite side of the Moon. Strange – we need to understand this better!

6: Mare Smythii

Nearly all lunar maria are named for weather conditions or states of mind, but this one honors a nineteenth-century English astronomer famous for cataloging double stars. The center of Mare Smythii is beyond the bottom left; here, we see the southwestern corner of the Smythii Basin, which holds one of the Moon's greatest concentrations of floor-fractured craters. Each of the large craters is shallow and has a concentric ring of ridges and fractures. The crater at the center-right edge, 37-km-wide Haldane, exhibits a steep-sided inner ridge, making it look like a medieval walled town. It is believed that each of these craters had under it a mass of magma that pushed up and fractured the crater's floor. But why was this activity so common in Mare Smythii?

7: Taruntius

Under full Moon lighting Taruntius has rays, so it is not ancient, like rayless craters such as Plato or Archimedes, yet it is not as spectacular as most young craters. Why? First, it is not very deep; a fresh crater with a diameter of 56 km is typically nearly 4 km deep, but Taruntius's depth is only 1.1 km. And its central peak actually rises higher than its rim – which is very unusual. An additional oddity is the concentric wreath of cracks and ridges on the floor. These are all characteristics of floor-fractured craters (FFC) that are thought to have originated as normal impact craters. All FFC are near maria, leading to the interpretation that magma rose underneath the craters, lifting their entire floors, with fractures marking the edges of the uplifted areas.

8: Mare Fecunditatis

Some maria fill deep basins and completely submerge preexisting features. Mare Fecunditatis (Sea of Fertility) seems to fill a more shallow depression, as inferred by the presence of outlines of earlier craters. Subtle ridges that can only be seen when the illumination is somewhat low trace the rims of such "ghost craters." The 12-km-wide sharp crater Ibn Battuta gives an idea of the diameters of these vanquished predecessors that were 20–30 km wide. Note that Ibn Battuta has almost no elevated rim, which means it formed on a lower surface of lava and that subsequent flows almost buried it. If those last flows had been any thicker, Ibn Battuta would have been inundated by lava and perhaps become a ghost itself.

9: Langrenus

Langrenus is a classic fresh impact crater, the Copernicus of the eastern limb. It has a central peak in the middle of a wide, partly flat floor, surrounded by terraces that formed as the walls collapsed during the last stage of crater formation. The edge of Langrenus's 127-km-wide rim is serrated where small, sub-circular collapses bit into a rim that was originally more circular. The crater is young enough to retain the smooth impact melt that coats the near side of the floor and fills low spots outside the rim – note the smooth patch in front of the crater. Over time, meteorite bombardment destroyed thin layers of melt and it ultimately disappeared. Along the bottom of the image is the old crater Lohse that was pummeled by ejecta from Langrenus.

10: Petavius

Petavius is almost a normal crater. Similar to nearby Langrenus (#9), Petavius has terraced walls and central peaks, but unlike any other crater on the Moon, the floor of Petavius is cut by a giant trench that connects its rim with the central peaks. Other smaller and more typical rilles mark Petavius as a floor-fractured crater whose floor was domed upward and fractured. Vertical views show that the trench bends sharply at the central peaks and continues as the slightly sinuous rille on the near side of the crater. Why Petavius has such a huge trench is unknown, but someday it will provide geologist-explorers with a magnificent cross section through the floor of a large impact crater.

If the Moon's interior had not heated up, producing volcanism, its surface would be much less interesting. But deep melting of the lunar mantle did happen, and magma rose to the surface to erupt as mare lavas, and also to seep under and sometimes onto the floors of craters. Near the margins of impact basins are some large craters that formed about the time mare lavas were flowing. Often these craters were invaded by magma, greatly altering their floors. Shown here is the 190-km-wide crater Humboldt near Mare Australe. The concentric and radial rilles are exactly what would happen if the center of Humboldt were pushed up from below. And that is what probably happened – magma filled the fracture zone under Humboldt, lifting up its floor. In a few places the dark lava broke through the crust and leaked onto Humboldt's floor. Humboldt is another floor-fractured crater.

12: Furnerius

The eastern edge of the Moon is loaded with large and mostly old craters. Ancient history is preserved in old craters such as Furnerius. This 135-km-wide crater preserves vestiges of its terraces, but the floor is mostly covered with debris from younger craters and perhaps from basins. A small patch of smooth lava marks the lowest part of the floor; a peculiar circumstance since there is no other mare material nearby. Stranger still, the floor is cut by a broad trough, but this does not mean that Furnerius is a floor-fractured crater. This broad rille, like the nearby Snellis Valley, is roughly radial to the Nectaris Basin, and its origin is probably tied somehow to that basin's formation.

13: Stevinus

Stevinus is one of the few large fresh craters near the eastern limb of the Moon. Like a smaller and less pristine Langrenus (#9), Stevinus has a terraced rim and a massive but off-center central peak. The floor seems to be covered with impact melt, but impact ponds are not visible beyond the rim. Perhaps Stevinus was a nearly vertical impact, sending its melted material straight up and back down, unlike an oblique impact that would have propelled melt in the downrange direction (e.g., Buys-Ballot, #88). The crater touching Stevinus on the bottom left side was severely affected by ejecta from the larger crater, but surprisingly did not have its rim pushed in. Sometimes impacts seem almost surgical.

14: Rheita Valley

Slashing across the southeastern section of the Moon is a 450-km-long valley. The large crater superimposed on the valley is the 42-km-wide Mallet D. The fact that it is battered emphasizes that the valley is old. In this lighting the valley north (top) of "D" appears more as a scarp, but below the crater its true nature is revealed as a series of touching craters. The key to undertanding the Rheita Valley is that it is radial to the Nectaris Basin hundreds of kilometers to the northwest. The valley is a secondary crater chain, just like the smaller ones seen at impact craters, but the projectiles ejected during the basin's formation were small mountains and the craters that make up the valley are up to 10 km wide – big secondaries. The Leuschner Chain (#80) illustrates what this might have looked like just after it formed.

15: *Kaguya* Impact Point

Some places of scientific or historic interest are not necessarily dramatic. This area of rounded terrain pocked by older craters is where the *Kaguya* spacecraft – which carried the TV camera that acquired all the images in this book – crashed onto the Moon. The collision was planned to end a mission that had brilliantly attained its goals. A spacecraft reaches the end of its mission when the fuel is nearly expended and the last bit is used to guide it into a location where the impact can be observed from Earth. Unique among lunar orbiters, *Kaguya*'s high-definition TV continued to obtain images as the spacecraft skimmed over the surface toward impact, and remarkably, astronomers in Australia imaged the bright flash as *Kaguya* struck the Moon's surface. The impact point was on the rim of this unnamed crater near Gill. Perhaps the crater should be named Kaguya!

16: Janssen Rille

Rilles are common on the Moon, except in the highlands. That is why Janssen is such an oddity. The crater lies deep in the highlands, away from mare deposits. Janssen is a large and confusing crater. Its actual size is uncertain because three or more craters overlap, including the youngest, Fabricius (*top right*), with its two ridges rather than a central peak. Janssen is probably about 180–200 km in diameter and is filled with ejecta. The rough-textured material in the top half of the crater, and perhaps the smoother surface at the bottom, is ejecta from the Nectaris Basin over the horizon to the north. The three rilles seen here cut the Nectaris ejecta but are infilled by debris from Fabricius. The rilles go in different directions, indicating that various forces acted to form them, but what caused those forces is unknown.

Impacts create craters, many small, a few huge. The biggest are called multi-ring impact basins, and the best preserved on the Moon is Orientale (#74). Probably second best – certainly the best preserved on the Earth-facing hemisphere – is the Nectaris Basin. The most conspicuous rim for this basin is so prominent that it has a name, the Altai Scarp. The segment of the rim curves around about 30% of the southwestern side of the basin, typically rising 1–1.5 km above the adjacent basin interior. To the right of the scarp are the ancient cratered highlands, and everything to the left is material created since the basin's formation, before 3.9 billion years ago. When the scarp formed by the collapse of the interior of the basin, it cut off a portion of the large flat-floored crater at *center-right*, resulting in the scarp being low.

18: Theophilus

Large fresh craters are the glory of the Moon. They have not yet been eroded and modified enough by subsequent small impacts to lose vital clues to their formation. The 110-km-wide Theophilus is a beautiful example of a complex impact crater. Below a steep rim crest – the slope was measured at 46° – a series of terraces step down 4 km to the floor below. A massive complex of peaks occupies the center of the floor, surrounded by blocks and mega-boulders that slid down the walls. Immediately north of the crater rim are numerous flat, smooth patches. These are ponds of impact melt – rocks made molten by the energy of the crater-forming impact – that came over the rim and collected in low spots. One more thing about the shape of Theophilus: It looks like a grand crater, but its depth is only 4% of its diameter; like all craters it is a very shallow depression!

Lunar mountain ranges are the rims of impact basins, but one of the mysteries is why some are so grand (e.g., the Altai Scarp # 17) while others are hardly noticed. The lunar Pyrenees belong to that latter category, marking the eastern edge of the mare in the Nectaris Basin. The rounded mountains run from the bottom of the image towards the top, rising only 1–2 km. They are far less spectacular than the 33-km-wide impact crater Bohnenberger, whose floor has risen and cracked like a freshly baked loaf of bread. This is an excellent analogy, for it is believed that a mass of magma rose up under the original floor of the crater, doming and fracturing it. This happened to a lesser degree to the similar-sized crater at the bottom of the image. These are floor-fractured craters.

20: Goclenius

The floors of many craters near the edges of the maria are flooded by mare lavas, and yet their rims are not broken. Under each crater is a deep nexus of fractures generated during its impact formation. Magma, rising up basin fractures, must intersect fractures under the craters and rise up those pathways to erupt onto their floors. In the case of Goclenius, the large crater in the center of this scene, the rising magma appears to have cracked the floor, creating the network of narrow rilles. This is the identifying characteristic of floor-fractured craters, but Goclenius also is fractured by other more regional stresses. The widest rille passes through the crater floor, cuts its rim, and keeps traveling 150–200 km further to the north. This long Goclenius Rille is roughly concentric to the Fecunditatis impact basin to the right and may have formed as the weight of the mare caused the center of the basin to collapse, cracking the marginal lavas.

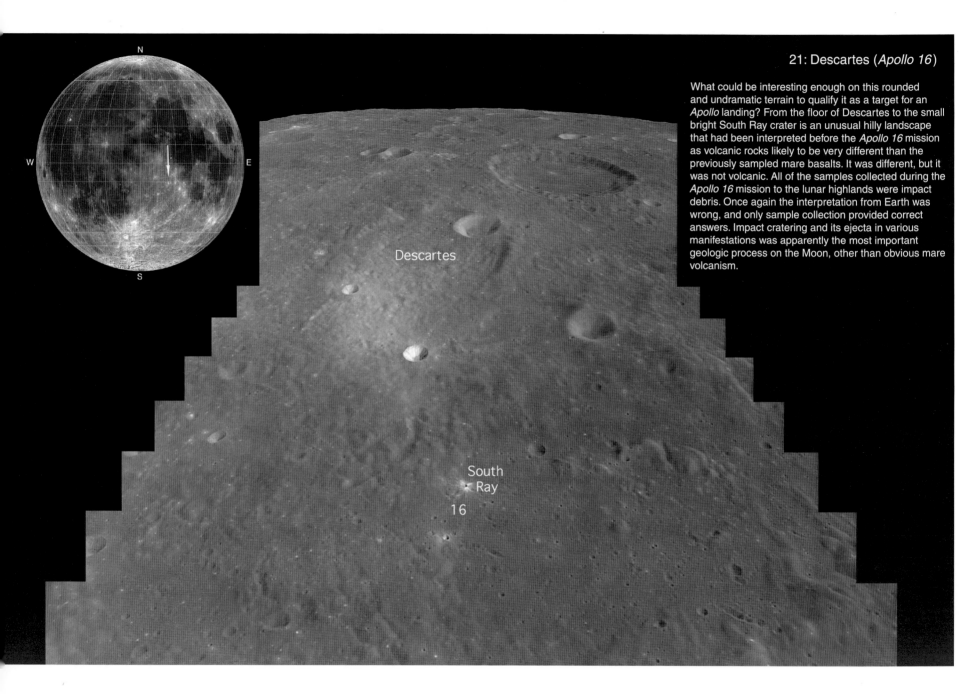

What could be interesting enough on this rounded and undramatic terrain to qualify it as a target for an *Apollo* landing? From the floor of Descartes to the small bright South Ray crater is an unusual hilly landscape that had been interpreted before the *Apollo 16* mission as volcanic rocks likely to be very different than the previously sampled mare basalts. It was different, but it was not volcanic. All of the samples collected during the *Apollo 16* mission to the lunar highlands were impact debris. Once again the interpretation from Earth was wrong, and only sample collection provided correct answers. Impact cratering and its ejecta in various manifestations was apparently the most important geologic process on the Moon, other than obvious mare volcanism.

Descartes

South
Ray

16

22: Tranquillity Base (*Apollo 11*)

"Houston. Tranquillity Base here. The Eagle has landed." These historic words were the first sent from the surface of the Moon. Neil Armstrong avoided unexpected boulder fields, and piloted the Eagle lander to a safe touchdown on the edge of Mare Tranquillitatis. The site had been selected for one main reason – safety – and yet it surprised us. The human scale details – boulders and crater pits that could wreck a spacecraft – became known only by being there. Similarly, the ages and compositions of lunar rocks could only become certain when samples were collected and returned to analytical laboratories on Earth. *Apollo 11* was the most audacious journey in human history, and because of it we learned the basics of the Moon, which became a foundation for understanding the Solar System.

Aldrin

Collins

Armstrong

11

Moltke

Plinius is the guardian of the strait between Mare Tranquillitatis (foreground) and Mare Serenitatis (near the horizon). Like
Ross at the bottom left, Plinius is a standard impact crater in the
transition range between smaller, simpler craters with unadorned
walls, and fully developed complex craters with terraces stair-
stepping down from the rim to the floor. Below steep cliff scarps,
both 24-km-wide Ross and 43-km Plinius have massive slumps
and only the hint of terraces. Beyond Plinius the boundary
between the maria Tranquillitatis and Serenitatis is quite sharp.
Lavas of different ages and compositions meet at the junction. The
darker lavas of Tranquillitatis and along the shore of Serenitatis are
older. Younger, light-hued lavas erupted into the center of
Serenitatis but were not voluminous enough to completely cover
the higher levels of the previous older flows near the basin rim.

24: Cauchy Fractures

The 12-km-diameter simple crater Cauchy is near the middle of this view, but few people point their telescopes at this region to see it. Both above and below Cauchy are remarkable fractures. To the south is the Cauchy Fault, a 120-km-long rupture in the lunar surface, where the north side moved up a maximum of about 350 m above the south side. North of the crater is the Cauchy Rille, which is a classic slightly curved trough, with a floor dropped down between two opposing faults. The rille is actually made up of three segments, with offsets at their ends. Both the rille and fault are approximately radial to the Serenitatis impact basin, so it is tempting to speculate that they are related to its formation, but that is unlikely. The basin formed long before these lavas in Mare Tranquillitatis, and the fractures occurred after the lavas cooled.

25: Tarus-Littrow (*Apollo 17*)

Dark-hued lava and bright mountains attracted mission planners to the *Apollo 17* site. The volcanic rilles and the darkness that could be due to pyroclastics or ash meant this could be an area with volcanic rocks younger than any sampled during the five previous *Apollo* missions. The surrounding bright mountains were probably part of the rim of the ancient Serenitatis impact basin and could provide a date for its formation. The *Apollo 17* lunar lander touched down in the Taurus-Littrow valley (marked 17), and in the course of conducting geologic fieldwork the astronauts discovered volcanic glass beads that were not young as hoped, but 3.6 billion years old. The rocks collected from the Serenitatis rim were also old – 3.9 billion years – but they had been expected to be perhaps half a billion years older. *Apollo 17* taught us that much of the activity of the Moon – intense impact cratering and widespread volcanism – was concentrated in the first billion years of its history.

26: Posidonius

On the eastern shore of Mare Serenitatis lies one of the
most fascinating craters on the Moon. About 95 km
wide, Posidonius has dissimilar halves. On the east
(*left*) a ridge curves inward from the rim, encompassing
a rille-crossed terrain that is tilted to the west. The
western part of Posidonius's floor is covered by smooth
lavas, and crossed by a most remarkable sinuous rille.
This lava channel starts inconspicuously near the
northern edge of the crater, and with amazingly tight
bends flows near the wall for about 50 km before
striking south across the flat floor. It then turns west and
absolutely hugs the crater wall, until disappearing at a
gap in the wall. Almost certainly the rille once emptied
lava onto the lower Mare Serenitatis, but no evidence of
this spectacular flow remains.

N

W   E

S

The 87-km-wide Atlas is 2 km greater in diameter than Tycho, but is only half as deep. The reason is clear. Instead of having a series of terraces that lead to a broad flat floor and a massive central peak, Atlas is shallow, with little flatness anywhere. Pieces of a possible central mountain are visible, but the entire floor is rough and cut by crudely concentric rilles. The floor on the near side and the far side is somewhat dark, suggestive of ash deposits. Altas started as a normal complex impact crater, but at some later time magma rose under it, lifting and cracking its floor. Some of the most gas-rich magma escaped, creating the dark halo ash deposits. Like Posidonius (#26) Atlas is a floor-fractured crater, but with a different manifestation.

28: North Pole

The scenery near the lunar north pole is not nearly as dramatic as at the south pole (#81), where the rim of the largest basin on the Moon gives rise to towering mountains. The north pole (just off the image to the *upper left*) is populated by craters, many rounded, filled in, and beaten down by ejecta from the formation of the Imbrium Basin. The north pole still is of great interest, because, like the opposite pole, the floors of some of its craters are in perpetual darkness and hence may hold water-ice. And the rim crests of some of its craters are almost always in sunlight. Light, for solar energy generation, and water, for drinking and rocket fuel, make the north pole a location of potentially great commercial value.

Plates 29 to 64

Lunar impact craters are like humans – they are all generally similar in form, but have many details that distinguish them from all the others. The 70-km-wide Philolaus on first glance looks like Anaxagoras, Tycho, and many other large, fresh craters, but its rim crest has deviations from circularity that give it a unique character. Each of these scallops marks where a section of the rim wall fractured off and cascaded down slope, making a piece of a terrace or simply a mass of mounded debris. Some boulders – actually small mountains a few kilometers wide – tumbled down the rim and halfway across the floor. The lack of an atmosphere means that no sound was produced when these gigantic landslides occurred, but many Moonquakes must have been produced, and clouds of dust must have created a temporary cloud over the crater.

This close-up image of the interior of the 50-km-diameter crater Anaxagoras was taken with the telephoto lens of the *Kaguya* HDTV. The resulting low oblique view clearly reveals the structure – both organized and chaotic – of the inner rim. To the left, the gentle rise of the outer slopes to the rim leads to a very sharp rim crest and then to a series of jumbled terraces, where masses of rock slid down toward the floor. Opposite, and partly in shadow, six different terraces indicate that portions of the original rim broke off and collapsed in a series of steps. On the far right, there are no terraces, just a 40° steep slope and piles of debris at the edge of the floor. We do not know why the inner walls collapsed in such different ways; perhaps it was related to weaknesses in the crust before the impact occurred.

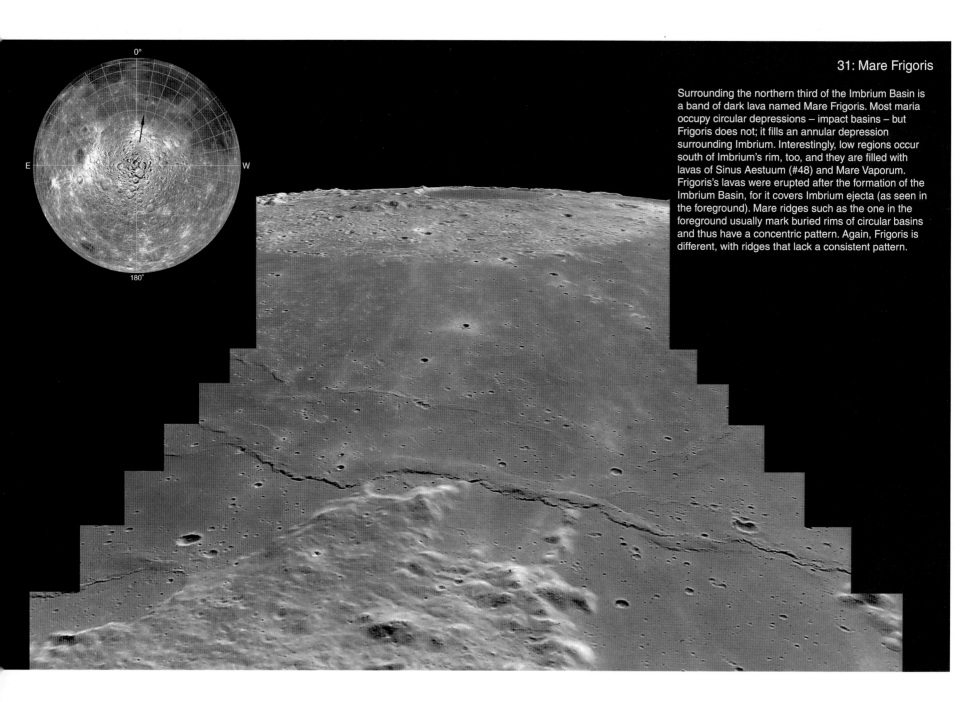

Surrounding the northern third of the Imbrium Basin is a band of dark lava named Mare Frigoris. Most maria occupy circular depressions – impact basins – but Frigoris does not; it fills an annular depression surrounding Imbrium. Interestingly, low regions occur south of Imbrium's rim, too, and they are filled with lavas of Sinus Aestuum (#48) and Mare Vaporum. Frigoris's lavas were erupted after the formation of the Imbrium Basin, for it covers Imbrium ejecta (as seen in the foreground). Mare ridges such as the one in the foreground usually mark buried rims of circular basins and thus have a concentric pattern. Again, Frigoris is different, with ridges that lack a consistent pattern.

32: Plato

One of the Moon's grand sights is the dark-floor crater of 101-km-wide Plato, just north of Mare Imbrium. In many views, Plato looks like a giant skating rink, but closer scrutiny reveals four or five craters 1–2 km in diameter and dozens of smaller ones. Lavas that erupted to form Mare Imbrium also created fissures under Plato, covering its original floor with about 2.5 km of lava that buried a bright central peak and wall terraces. Plato is famous for being the location of numerous reports of mists, fogs, and other transient lunar phenomena. None of these has ever been confirmed, and Plato, like everywhere else on the Moon, is unchanging except for the occasional impact of solar system debris creating new craters a few tens of meters wide.

33: Alpine Valley

The Moon has a mountain range called the Alps, but they are not ice-capped peaks jutting into a deep blue sky. The lunar Alps are mountain-size boulders of ejecta from the formation of the nearby Imbrium impact basin. About 3.8 billion years ago a rain of mountains devastated and buried preexisting terrain, leaving behind a jumbled field of debris. Cutting through this is a 165-km-long and roughly 20-km-wide trough that is radial in the center of the Imbrium Basin. This Alpine Valley has a flat floor with a tightly meandering rille running down its middle. The trough does not appear to have been formed by its spreading apart, because the opposite sides do not match. It must have formed after the ejecta was deposited, with the center down-dropped so far that all the mountainous debris was buried under lavas that welled up the fractures. It would be misleading to say that scientists completely understand what happened here.

34: Aristoteles

The large crater Aristoteles illustrates two oddities. First, its diameter of 87 km, nearly the same as another fresh crater, Tycho (#44), implies that it should have broad terraced walls and massive central peaks. The terraces are there, but there are only little runt peaks in a broad, fairly smooth floor. Tycho is 1.2 km deeper than Aristoteles, implying that its floor has been made more shallow by something. It could be volcanism but more likely is a thick deposit of impact melt. In fact, streams of melt are visible that flowed from the near rim crest, 35 km down slope. The second oddity about Aristoteles is that its rim sliced through a smaller crater, Mitchell. In general, larger craters formed early in lunar history and became less frequen5 with time. That is why it is more common to see small craters impinging on bigger ones. But Aristoteles is an intriguing exception.

Archimedes reminds us of Copernicus, a fresh complex crater with terraced walls and central peaks on its floor. The walls and terraces of Archimedes are similar to those of Copernicus, but the floor is much wider and there are no central peaks. Because of these differences students of the Moon once considered Archimedes and Copernicus to be two different types of craters. But simple measurements and a thought experiment suggest an alternative. Notice that the floor of Archimedes is the same dark-hued, ridged material as the mare that surrounds it. Imagine that Archimedes was flooded by mare lavas that rose up fractures that underlie impact craters. Comparison of the depth of Copernicus and the heights of its central peaks with similar measurements for Archimedes implies that the lavas completely bury Archimedes' peaks, transforming what must have originally looked like Copernicus into the feature we see today.

36: Hadley Rille (*Apollo 15*)

Hadley Rille is justifiably famous for being one of the most dramatic landing sites for *Apollo* astronauts. The crew swooped in low over the towering Apennine Mountains (seen in the *bottom right*) and dropped down precisely between the front of the mountains and the sinuous Hadley Rille. The landing site was selected so that the astronauts could easily sample rocks from the Apennines – the rim of the giant Imbrium Basin – and look into the rille to study the layers making up the Imbrium lava flows. This region is doubly important historically, because the Soviet *Luna 2* probe, the very first object from Earth to touch the Moon, crashed somewhere within the area shown at the *upper left*.

The vast majority of volcanism on the Moon consisted of eruptions of sheets of lava that made deep piles of maria within impact basins. Some of these great basins took more than a billion years to fill with lavas, whose composition and character changed over time. A graphic example occurs on the southwestern shore of Mare Serenitatis, near the simple crater Sulpicius Gallus (*bottom right*). The main mare is relatively smooth and crossed by ridges. However, between the edge of the mare and the hilly basin rim is a darker and rille-cut material. Because the smooth mare lavas embay and truncate the material with rilles, the latter must be older. The darkness is because these older lavas are draped by pyroclastics or volcanic ash that was erupted from the rilles. Orbiting astronauts noted that these ash deposits had flecks of orange, probably from orange volcanic beads such as had been sampled across Mare Serenitatis at Hadley Rille (#36).

38: Hyginus Rille

Hyginus is one of the few craters on the Moon that is unlikely to have formed by impact. It does not have a raised rim, and it is in the middle of a 100-km-wide gentle depression about 1 km deep centered in the crater and its rilles. Hyginus is at the junction of two major rilles and two faint ones. The floor of Hyginus is covered with small bumps, very unlike floors of normal impact craters, and the crater seems to truncate the rilles – to be younger and to cut through them. The major rilles extending from Hyginus are unique to the Moon, but are familiar to geologists who have worked in Hawaii. The 3-km-wide flat-floored rilles are pierced by more than a dozen collapse pits, clearly visible in the rille at right front. In Hawaii, a chain of collapse pits – without the bounding rille – forms over a major subsurface conduit that distributes lava that rises under a central caldera to its flanks. Was Hyginus a caldera and feeder for massive eruptions?

The 26-km wide fresh crater Triesnecker is the guidepost for a "K"-shaped pattern of rilles. The rilles were there first, because ejecta from the crater fill the rille segment nearest to it. That there are different generations of rilles is demonstrated by the observation that the straight leg of the "K" has softer edges and thus is older than the other parts. Between the rilles near Triesnecker and the peculiar Hyginus Rille (#38) in the distance is another group. Although these rilles are at a rough 45° angle to the main "K"-shaped ones, there seems to be some tenuous connections. Linear rilles like these are typically surface expressions of subsurface dikes of rising magma. Such dikes and their rilles can form only if there is extension in directions perpendicular to them. The 45° rilles are roughly radial to the Imbrium impact basin and may have formed as a result of forces associated with its formation, but it is unclear what caused the "K" rilles to form.

40: Ariadaeus Rille

Rilles are troughs that cut the lunar surface. Most traverse mare and are related to impact basins or volcanism. The Ariadaeus Rille is large and easy to see telescopically, but hard to understand because it is so atypical. It is big – 220 km long, 4.5 km wide, and about 500 m deep. Similar, but generally smaller, troughs occur on Earth, where they are called graben. The floor of a graben – or rille – slides down between two parallel faults. This can only happen if there is extension or moving apart of the crust to allow space for the floor to collapse into. The best guess for the reason for expansion around Ariadaeus is that underneath is a volcanic dike, a vertical sheet of magma that slightly raised the ground over it, cracking the surface.

One thinks of craters when thinking of the Moon, and the nearside highlands are full of craters. Each was formed by a random impact of an asteroid or a comet, and through modification assumes a unique morphology. Alphonsus is the floor-fractured crater occupying the bottom of this image. It is 108 km wide and has a broad smooth floor within a rim raked by ejecta from the formation of the Imbrium impact basin. Alphonsus is famous for small dark halo craters that lie on narrow rilles along the edges of its floor. The rilles and small craters are volcanic features, proving that lavas can colonize impact craters. Arzachel, the fresher crater above Alphonsus, is also a floor-fractured crater but lacks volcanic pits, having only roughly concentric rilles, one of which casts a shadow, showing that it is also a fault. Alpetragius is a different type of crater. It is the odd one on the right with a smoothed interior and a giant mound of a central peak. One astronomer compared Alpetragius to a bird's nest containing a single giant egg.

42: Straight Wall

The lunar surface is broken by the impact of projectiles that have excavated craters. But only rarely is it broken by other processes. Here is the best example of a lunar fault. The Straight Wall (or Rupes Recta, to use its official Latin name) stretches 110 km across the floor of an ancient crater whose western (*left*) side has disappeared beneath the encroaching lavas of Mare Nubium. In this picture the Sun is to the east (*right*), casting shadows into Birt, the 17 km diameter crater west of the Wall. The Straight Wall itself casts a shadow, too, showing that its right side is about 500 m higher than the left. A smaller curved fracture is to the west of Birt. It is a small volcanic rille with collapse pits at both ends.

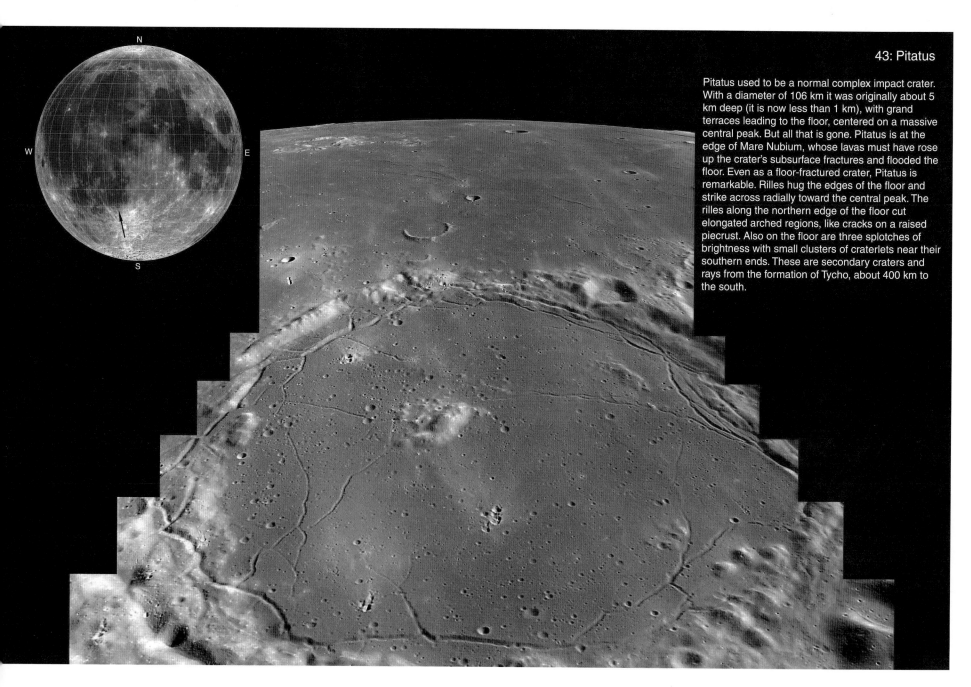

Pitatus used to be a normal complex impact crater. With a diameter of 106 km it was originally about 5 km deep (it is now less than 1 km), with grand terraces leading to the floor, centered on a massive central peak. But all that is gone. Pitatus is at the edge of Mare Nubium, whose lavas must have rose up the crater's subsurface fractures and flooded the floor. Even as a floor-fractured crater, Pitatus is remarkable. Rilles hug the edges of the floor and strike across radially toward the central peak. The rilles along the northern edge of the floor cut elongated arched regions, like cracks on a raised piecrust. Also on the floor are three splotches of brightness with small clusters of craterlets near their southern ends. These are secondary craters and rays from the formation of Tycho, about 400 km to the south.

44: Tycho

Tycho is one of the most spectacular craters on the Moon because it is young, at least by lunar standards. Considering that most craters and mare lavas are at least 3 billion years old, Tycho's youthful 100-million-year age is just yesterday. The reason Tycho is still fresh is that few large craters and no significant lava flows have occurred in the last billion years of lunar history. And since the main erosion acting on the waterless and airless Moon is due to the creation of impact craters, the paucity of new craters means that little has changed in Tycho. But you can see the damage its formation did – pasty ejecta from Tycho instantly aged preexisting craters. All the foreground craters were coated by melted and pulverized rock fragments that swept across the surface by Tycho's impact. A beautiful, pristine crater rains destruction on its predecessors.

This panoramic view from rim to rim across 83-km-wide Tycho shows what most large impact craters originally looked like. Wall terraces, flat floor, and central mountain are common to all large complex impact craters, but also preserved here is the patina of impact melt that coated the wall and pooled on the floor. During an impact event the target rocks that are most intensely affected by the energy are in fact completely melted and ejected out of the newly formed crater. The material falls back still in a molten state and flows downhill, until it pools or solidifies. The fact that impact melt drapes the terraces proves that they form during the impact, not later as an adjustment. The spidery black web on Tycho's floor is made of shadow-filled cracks caused by contraction of the cooling melt sheet.

46: Bullialdus

Even though it looks like a slightly smoothed Tycho, Bullialdus is a magnificent sight. Below its steep rim scarp, four or more terraces step to the broad floor, with a massive peak complex in the center. The half dozen or so conspicuous craters within Bullialdus demonstrate that it is older than crater-free Tycho. Surrounding Bullialdus is a hilly terrain, giving way in the middle distance to chains of small secondary craters. Just beyond Bullialdus's rim is a crater – Bullialdus A – that is softened, and beyond that is a slightly smaller one – Bullialdus B – that is very sharp-rimmed. A existed before Bullialdus's formation and was pelted and veneered with ejecta, accounting for its subdued look. B is superposed on top of ejecta from Bullialdus and thus must be younger. Beyond B is an older crater, Kies, which is older than the Mare Nubium lavas into which Bullialdus impacted. Mare lavas surrounded and filled Kies, converting it into a ruin.

47: Fra Mauro (*Apollo 14*)

Apollo landing sites were carefully selected, requiring them to be of maximum scientific interest yet high in safety features. Fra Mauro was selected because it is a broad field of ejecta from the Imbrium Basin; collecting samples would provide the age of the basin-forming event and evidence of the chemistry of the deep lunar crust. The Fra Mauro crater (lower center) is 100 km wide, but so ancient that it was flooded by lava, crossed by rilles, rained on by Imbrium ejecta, and pelted by random impacts. Nearly all of the samples collected were breccias – fragments of rocks broken by one or more impact events. The age of the Imbrium Basin was dated at 3.85, establishing a major time horizon. With that date, many craters across the Moon that had been previously established by geologic relations as pre- or post-Imbrium now were tied to an absolute age. Small fragments of volcanic rocks were found in the breccias, including some that were 4.2 billion years old, establishing that volcanism had started early in lunar history.

48: Sinus Aestuum

Sinus Aestuum is the circular patch of mare lava partially seen on the right; it probably fills a small impact basin. Lunar scientists are more interested in the dark area in the center of the image that is cut by the straight rille named Bode II. The dark area is believed to be a thin dusting of volcanic ash or pyroclastic material, probably erupted from the rille. The dark mantle is thought to be due to volcanic glass beads rich in the elements iron and titanium. The ash probably erupted in gas-rich fire fountains, with lava shooting up thousands of meters into space and falling back down to coat the surface. Of course, the mystery here is what gas propelled the pyroclastics. On Earth water vapor drives most eruptions, but lunar rocks contain almost no water, so other gases such as carbon dioxide or sulfur may have provided the motive force. Note the surface at upper right, which is another expansive pyroclastic deposit. This area of the Moon was remarkable in hosting so much explosive volcanism.

Copernicus is a typical example of all large craters on the Moon. Unlike small, simple, bowl-shaped craters, Copernicus has a complex interior. Its rim was constructed from fallback of ejecta from the giant impact that excavated it. Additionally, rocks at the edges of the hole rebounded upward following the impact, and as they rose, massive rings of material broke off and slid toward the floor. Some of the material moved down as coherent blocks, creating terraces that can be traced nearly all the way around the rim. And directly under the point of impact the floor also rebounded, pushing up mountains of material originally 5-10 km below the surface. The flat floor is covered with rubble from the walls and fallback ejecta, and in the upper left, especially, there is smooth material made of rocks that were melted by the energy of impact and launched into space, later splashing back down to the surface.

The Kaguya Lunar Atlas

50: Copernicus Secondary Craters

Large impacts such as the one forming Copernicus (#49) not only dig impressive holes but also scatter the excavated material near and far; some of this material probably reaches Earth as meteorites. Close to the impact site, ejecta fall back, helping to construct the crater's elevated rim. Further away, continuous blankets of ejecta give way to a rain of individual particles and clusters of debris that make secondary impact craters. These secondaries east of Copernicus are a few kilometers across and occur in lines, ribbons, and randomly. Often the pits overlap, indicating that the clot of debris that formed them broke into pieces before hitting the ground. The area in the foreground is similar in composition and age to the Mare Imbrium lavas near the horizon, but has been pulverized and lightened in hue by the rain of secondary craters.

Seen from Earth, the lunar maria look dark, flat, and relatively featureless. But from lunar orbit, details emerge that document their histories. This south-looking view shows the northern part of Mare Insularum (Sea of Islands) near the sharp-rimmed crater Milichius at upper left. Two types of volcanic features are visible – rilles and domes. The narrow rille at bottom center is typical of appropriately called sinuous rilles. They wind across the surface with large and small bends. Just to the left of the bottom of the rille is a very gentle rise with an elongated pit near its summit. This 100-m-high landform is a volcanic dome. Another dome is to the right of Milichius. This one, called Milichius Pi, is about 10 km in diameter and 230 m tall, with a small pit on its top. Domes result from eruptions of lavas that do not flow far, building up low mountains around their vents – the pits.

Rilles are channels that carry lava across maria, and the bottom one curved around the preexisting dome.

52: Kepler and Encke

Rubble, craters, and mare comprise this view. The lumpy hills are ejecta from the formation of the Imbrium Basin over the horizon, and later lavas of Oceanus Procellarum infiltrated low spots. Kepler, the fresh-looking crater above center, and Encke (*bottom right*) are both about 30 km in diameter but look very different. Kepler is a standard small impact crater with complex morphology: wall slumps and incipient terraces, flat floor, and central peaks. Encke may have originally looked nearly the same, but now it is a floor-fractured crater, with ridges and rilles circling a central, less-disturbed area. Because the Encke-forming impact occurred when magma was still available beneath the surface, it rose up the crater's fracture zones, distorting and modifying the original morphology. Kepler formed later, when no magma was available at depth, so it retains its pristine shape.

53: Gruithuisen Domes

Seventeen percent of the nearside of the Moon are covered by mare lava flows, often studded with domes, sinuous rilles, and dark halo craters. These two steep-sided hills, named after the flat-floored crater Gruithuisen near the middle of the image, are different from nearly all of the Moon's other volcanic landforms. The fact that they are steep differs from mare domes, which have slopes of only a degree or two. These domes are not dark, like most volcanic features, but are bright, and they also are brighter in the red part of the spectrum than normal lunar lavas. All these characteristics suggest that the domes have a different chemical composition, and that their lavas had a high viscosity and did not flow far from the vents. Why did a different type of lava erupt in this part of the Moon?

A giant bay deforms the northern curvature of Mare Imbrium. Although the incursion of mare lavas is named Sinus Iridum, the more important feature is the range called the Jura Mountains that circle the northern half of the bay. The Jura are the remaining rim of a 260-km-wide impact crater; the question is, what happened to its missing southern half? Because the Jura stop abruptly at Heraclides Promontory (*bottom center*), rather than slowly decreasing in elevation, it is unlikely that they continue under the mare lavas. More likely, faults along the edge of the Imbrium Basin dropped down the southern rim a kilometer or more. Before the flooding by Imbrium lavas this would have been a remarkable sight, a huge crater chopped in two!

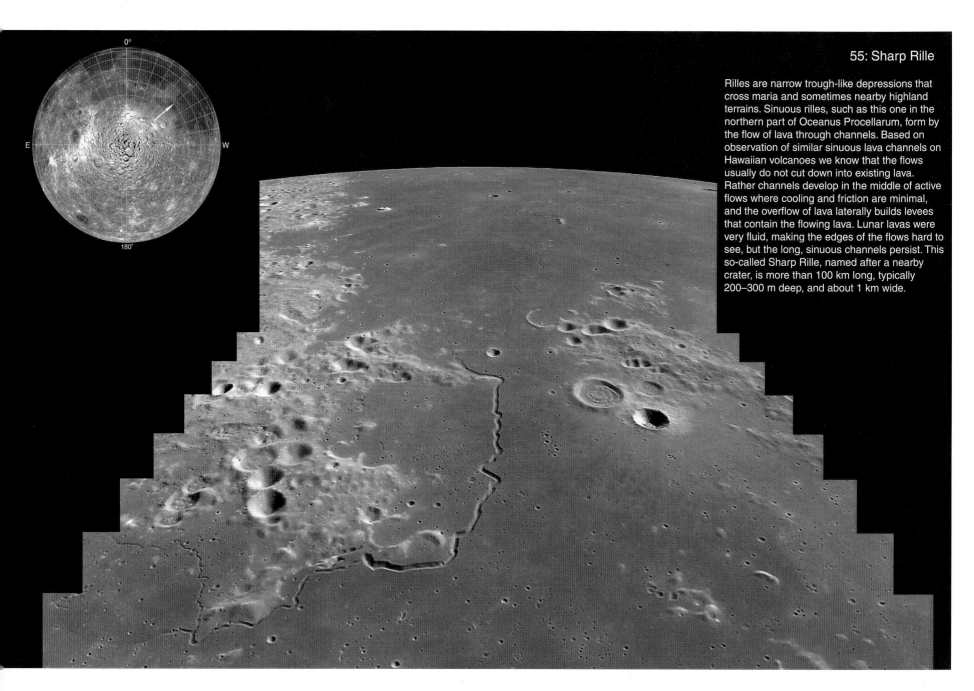

55: Sharp Rille

Rilles are narrow trough-like depressions that cross maria and sometimes nearby highland terrains. Sinuous rilles, such as this one in the northern part of Oceanus Procellarum, form by the flow of lava through channels. Based on observation of similar sinuous lava channels on Hawaiian volcanoes we know that the flows usually do not cut down into existing lava. Rather channels develop in the middle of active flows where cooling and friction are minimal, and the overflow of lava laterally builds levees that contain the flowing lava. Lunar lavas were very fluid, making the edges of the flows hard to see, but the long, sinuous channels persist. This so-called Sharp Rille, named after a nearby crater, is more than 100 km long, typically 200–300 m deep, and about 1 km wide.

56: Procellarum Ridges

Maria often look flat, but when the illumination is grazing, long ridges often become visible. The fact that mare ridges are hard to see unless the Sun is low means that their average slopes are only a few degrees. These ridges in northern Oceanus Procellarum display characteristic morphology: a broad, gently rounded base, surmounted by a narrower and steeper ridge. The mare ridge at bottom right is about 7 km across, with the steep part only 1 km wide, and probably only tens of meters high. Mare ridges used to be called *wrinkle ridges* because they look like wrinkled cloth. The modern interpretation is that they are shallow faults formed as solidified mare lavas collapsed into a basin, forcing outer parts to fracture and slide over inner parts. They are not folded, but one section of mare does overly adjacent pieces.

57: Rümker

Like a lumpy pancake, the broad volcanic dome Rümker sits on the surface of northern Oceanus Procellarum. Most domes are much smaller – Rümker is 65 km wide – and have more of a flattened hemispherical shape. In fact, a couple of such small domes occur near the top part of Rümker. Because its surface is pitted by a larger number of small impact craters, we know that Rümker is older than the surrounding Procellarum mare lavas. Rümker itself is made of material of two different ages. The broad circular area on the left has a smoother and less cratered – and therefore younger – surface than the edges of Rümker. Scientists do not know why Rümker is so much larger than other domes.

The Kaguya Lunar Atlas

58: Lavoisier

The western shore of Oceanus Procellarum has been an area of extensive lava leaks. Magma rose up fractures under craters, sometimes erupting smooth dark flows onto their floors, more often pooling under the floor and bodily lifting it. This produced doming and fracturing of the floor, creating floor-fractured craters. Lavoisier, a 70-km-wide much modified crater has a circle of hills interspersed with rilles in a wreath around its floor. In the middle of the floor is an odd and unusual landform called a concentric crater, named for the rim-within-a-rim structure. These largely occur in areas of volcanism, and the inner rim may be a lava extrusion onto the floor of an impact crater. The floor of Lavoisier F has been steeply domed and cracked by the intrusion under it, whereas "H" looks like its floor has collapsed – like a fallen pie crust – producing a random pattern of cracks. Lavoisier E has similar rilles and hills as Lavoisier, and "B," which is deeper and thus had less uplift, has a correspondingly smaller pattern of fractures.

Prinz is just another remnant crater, its south wall missing and its interior flooded by Oceanus Procellarum lavas. But Prinz and the nearby Harbinger Mountains are on a local high spot, probably elevated by an infusion of magma below, that fed four spectacular sinuous rilles. Each starts in an oval rimless depression and meanders down slope. The large depression to the right of Prinz has the beginning of a sinuous rille, but the rest of it must have been buried by later lavas. Similar lava channels in Hawaii also often have an enlarged pit over their vents, making a dramatic show at night when a pool of incandescent lava lights the sky, while a river of lava flows away.

60: Aristarchus Plateau

Surrounded by lavas of Oceanus Procellarum, the Aristarchus Plateau – named for the fresh crater on the right – stands out as older, higher, rougher, and delicately tinted. It is older because it has more impact craters that the surrounding mare, which also laps up against it. Measurements show that the plateau rises up to 2 km above the mare and contains hills, ridges, and valleys, most spectacularly Schröter's Valley, a giant sinuous rille that was the source for some of the Oceanus Procellarum lavas. The entire plateau is draped with 10–30 cm thickness of volcanic ash that sometimes gives it a delicate pink or yellow hue when observed with a telescope. Like the nearby Prinz (#59) area, the Aristarchus Plateau has been uplifted by a massive infusion of magma beneath it.

61: Marius Hills

Lunar volcanism was mostly limited to mare lava flows and accompanying sinuous rilles and mare domes. A 200-km-wide cluster of hills west of the crater Marius is a unique aberration. About 300 hills, typically 5-10 km wide and a few hundred meters high clump together like spilled raisins. A few sinuous rilles cut through the hills, and as shown here significant low flows of lava occurred, too. But unlike nearly all other volcanic cones on the Moon, the Marius Hills are relatively steep-sided and rough-textured. Studies of reflectance of the hills indicate that they are made of lava flows capped with volcanic ash. On Earth such features typically occur when there is enough entrained gas – mostly water vapor – to shred rising magma into ash particles. But the Moon had little water to convert magma to ash, so some other volatile – perhaps carbon dioxide – must have been richly available in the Marius area to build hundreds of ashy cones.

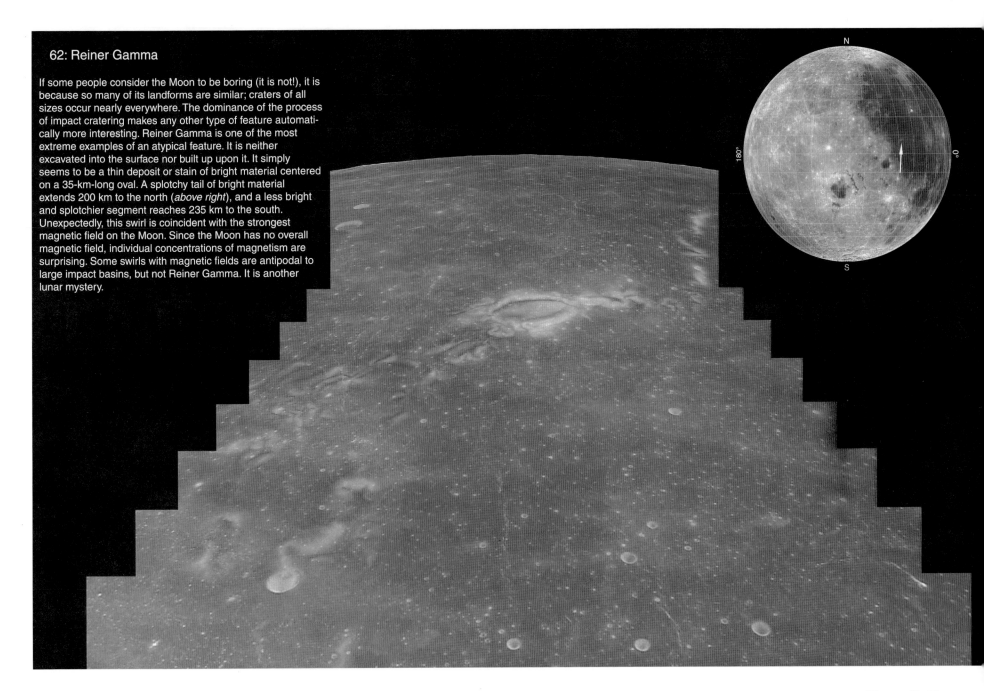

62: Reiner Gamma

If some people consider the Moon to be boring (it is not!), it is because so many of its landforms are similar; craters of all sizes occur nearly everywhere. The dominance of the process of impact cratering makes any other type of feature automatically more interesting. Reiner Gamma is one of the most extreme examples of an atypical feature. It is neither excavated into the surface nor built up upon it. It simply seems to be a thin deposit or stain of bright material centered on a 35-km-long oval. A splotchy tail of bright material extends 200 km to the north (*above right*), and a less bright and splotchier segment reaches 235 km to the south. Unexpectedly, this swirl is coincident with the strongest magnetic field on the Moon. Since the Moon has no overall magnetic field, individual concentrations of magnetism are surprising. Some swirls with magnetic fields are antipodal to large impact basins, but not Reiner Gamma. It is another lunar mystery.

The 42-km-wide crater Sirsalis at top-center is famous for the rille that passes near it. The Sirsalis Rille stretches 425 km across the lunar countryside, the longest rille on the Moon. Its average width and depth are 3.5 km and 230 m, respectively. It is a simple trough or graben, with the floor dropped down between two facing faults. Graben are depressions formed over vertical ribbons of magma called dikes. Dikes rise through crust where the horizontal stress is extensional, making it easier for the rising magma to push aside the surrounding rocks. Mathematical models suggest that the Sirsalis dike extended from deep within the Moon – about 300 km – to within 2.5 km of the surface. It is likely that the dike transported magma from its source region to erupt and produce the Oceanus Procellarum mare lavas.

64: Gassendi

Gassendi formed on the edge of the Humorum Basin and tilts toward the basin center. Lavas from the same source as those in Mare Humorum rose up basin fractures, and flooded a low-lying annulus inside the crater's southern rim. (Notice the narrow mare ridges inside the rim that seem to continue the one outside at bottom right.) With a diameter of 101 km, Gassendi was originally about 4–5 km deep, but now the floor is only 1.4 km below the rim crest. The floor was probably elevated by magma rising from below, fracturing the already hardened crust, creating the rilles that cross its floor. Gassendi is a floor-fractured crater. The younger crater that cuts the northern rim of Gassendi is a normal impact crater, but because it formed on a slope its terraces are more extensive on the high distant side, and almost nonexistent on the lower part inside Gassendi.

Plates 65 to 100

65: Hippalus Rilles

Like train tracks receding into the distance, three troughs curve around the edge of the Humorum impact basin, whose gray lavas are just visible to the left. As mare lavas filled the giant basin their weight caused its floor to sag. Bending along the basin's edge pulled apart the crust, creating the three rilles. The central collapse also tilted Hippalus crater (diameter 58 km) so that mare lavas flowed over its western rim, burying half of its floor. A much smaller rille cuts the lava-filled floor of the 48-km-wide crater Campanus at bottom right. Its crater walls are unbroken, so the magma from the lunar mantle must have risen up through fractures under the crater to erupt onto its floor.

66: Liebig Fault and Doppelmayer Rilles

Near the bottom of this image you can see the lavas of Mare Humorum lapped up against a gentle shoreline, but further north a fault sharply bounded their extent. That the original land to the right of the fault moved downward is proven by the half crater cut by the fault – its right side is buried under the lavas of Mare Humorum. Just below is a sharp-rimmed simple crater, Liebig E, which cuts the older terrain to the left, the fault, and the mare itself. A long linear rille and subsidiary ones – the Doppelmayer Rilles – slice through the mare. Under high Sun illumination the area around the rille is dark, suggesting that pyroclastics, or volcanic ash, erupted from it. The elongated depressions at the bottom right end of the rille indicate that it formed by collapses, perhaps into voids created by eruption of the ash.

E

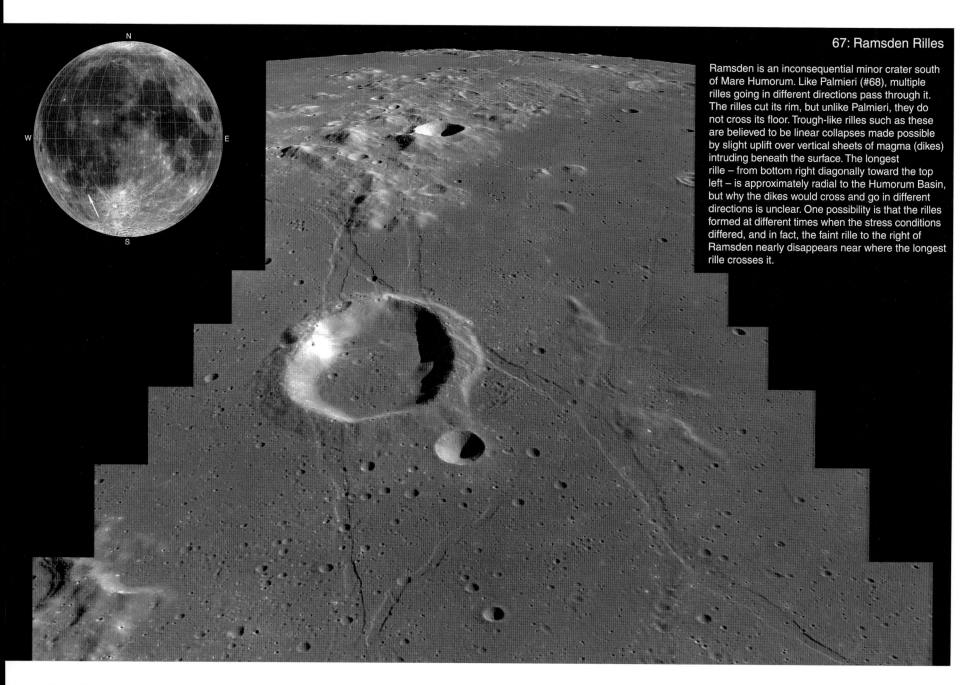

Ramsden is an inconsequential minor crater south of Mare Humorum. Like Palmieri (#68), multiple rilles going in different directions pass through it. The rilles cut its rim, but unlike Palmieri, they do not cross its floor. Trough-like rilles such as these are believed to be linear collapses made possible by slight uplift over vertical sheets of magma (dikes) intruding beneath the surface. The longest rille – from bottom right diagonally toward the top left – is approximately radial to the Humorum Basin, but why the dikes would cross and go in different directions is unclear. One possibility is that the rilles formed at different times when the stress conditions differed, and in fact, the faint rille to the right of Ramsden nearly disappears near where the longest rille crosses it.

The Kaguya Lunar Atlas

68: Palmieri

The area west of Mare Humorum is full of linear rilles.
The reason is not obvious, for although some of the
rilles seem to be related to the Humorum impact basin,
others go their own way without any obvious controlling
structure. The crater Palmieri is at the intersection of
two rille families. One broad rille – about 2.5 km wide –
trends to the north-northeast, curving in to the Liebig
Fault (# 66) that bounds Mare Humorum. Crossing this
rille at about a 60° angle is another flat-floored one
that is roughly tangential to the Humorum Basin. The
fact that these rilles cross mare material and highlands
means that they are more recent than either. The
difficulty in seeing the highland segments of the rilles
suggests that the highland material is less coherent
than the lavas and cannot support the step rille walls.

Schickard is a large, easy to identify crater because of its zebra stripe. The crater is 227 km in diameter, with a wide floor filled with relatively smooth material. The dark stuff on the ends is mare basalt, but what about the lighter stripe across the middle? It is ejecta from the formation of the nearby Orientale Basin, and a few of the craters on it are surrounded by dark halos, which are ejecta of underlying mare basalts. But the mare patches at either end of Schickard are younger than the stripe because they contain fewer subsequent impact craters than are on the stripe. So the sequence of events is: Schickard formed and is flooded with mare basalts. The Orientale Basin formation sent a swath of light ejecta over part of Schickard's floor, including the secondary craters and ridges of ejecta that modified the near floor. Later, additional mare lavas erupted at the ends of the crater.

70: Wargentin

This is one of the weirdest craters on the Moon. Wargentin is famous for not having a depression within its rim, but instead being filled to overflowing (*top left*) with some smooth surface material. It was long assumed that lava flows filled the original crater that must have been about 4.5 km deep – as is Tycho, which has the same diameter of about 85 km. Just northwest of Wargentin is the huge Orientale impact basin that may well have filled the crater with its ejecta. But some of the small craters on Wargentin are surrounded by dark halos, which have been determined to be mare basalt. So the material filling the floor is mare lava flows that were lightened by a veneer of bright ejecta from the formation of Orientale.

Impact craters are round because they are essentially point source explosions that throw material out in all directions. So what is wrong with Schiller, a structure 180 km long but only about 70 km wide? It appears to be made of three or more overlapping circular craters whose shared walls are missing. The raised rim and smooth floor of Schiller makes it look like an impact crater, but the lozenge shape requires explanation. Although earlier students of the Moon proposed that Schiller was a gigantic volcanic caldera, there is no evidence for that. A clue comes from the central ridge at the upper end of the structure and the coalescence of multiple craters. Similar features are seen in very low angle oblique impacts created in NASA's laboratories. Most real ones on the Moon are relatively small – Schiller is simply a giant example. Perhaps a small asteroid or comet was captured into lunar orbit and while spiraling inward, was torn into multiple pieces with the final near-grazing simultaneous impacts creating overlapping craters.

72: Drygalski

The interiors of large impact craters often have smooth floors of unknown composition. In some cases such as Plato (#32), the craters are near maria, and like maria have floors that are dark when seen under high Sun conditions, so it is easy to conclude that such floors are mare lavas. Other smooth floors are in young craters with impact melts outside their walls (e.g., Theophilus #18), and it is likely that the floor material is melt, too. Drygalski represents a different case, though, for it is near the south pole, far from any mare lavas, and it is an older crater and no impact melt has been reported. Like Drygalski, many craters in the southern highlands of the Moon have smooth, light-hued floors. Perhaps this floor material is a type of volcanism that was not sampled during the *Apollo* missions. Most scientists do not think that is true, however, because there are no volcanic landforms visible. The reigning opinion is that the smooth material is fluidized ejecta from the formation of the Orientale Basin about 3.8 billion years ago.

73: Hausen

Low oblique images taken with the HDTV telephoto lens make us feel like we are flying over the landscape. In this image the space-craft was only 23 km above the southern rim of the crater Hausen, looking across its floor and central peaks to the terraced northern rim. This low perspective emphasizes how shallow lunar craters are compared to their diameters. Hausen is 6 km from rim crest to floor, making it one of the deepest craters on the Moon. But compared to its diameter of 167 km, the depth is only about 4% of its width. Sometimes we think of impact craters as being deep structures, like teacups, but really they are more like the saucers.

74: Orientale Basin

Impact basins are the largest landforms on the Moon, and this image looks across the youngest. At bottom are the Cordillera Mountains (#75), the topographic rim of the basin, and the dark material near the top is the small lava pond that is Mare Orientale. Between the Cordillera and the mare are two different terrains. Nearest the mountains is a hummocky mass of small hills embayed with somewhat smooth material. This is the actual rim of the basin, the Outer Rook Mountains. Bright peaks of the Rook and the Cordillera on the opposite side of the basin are visible beyond the mare. Between the Rook and the mare is a lower undulating surface with small cracks, and closer to the mare, larger ones. This is impact melt, full of cracks that developed as it cooled.

75: Cordillera Mountains

The 930-km diameter rim surrounding the Orientale Basin is called a mountain, but it is really a scarp. Like the Altai Scarp (#17) enclosing part of the Nectaris Basin, the Cordillera rises gently above the distant surrounding plains, but it drops down precipitously to an interior bench on the left. Other than the ribbons of dark mare basalt that rose up the fault, the surfaces on both sides of the scarp are thought to be basin ejecta. This means that the actual rim of the Orientale Basin is interior to this scarp. Someday an astronaut will stand at the top of the scarp and wonder how he or she will reach the basin floor.

76: Orientale Mare

This strange scene is unlike any other mare surface on the Moon. Orientale was the last large impact basin formed on the Moon, and it has a relatively thin coating of mare material. This piece of Mare Orientale is typical of the rest. It is cut by numerous channels and has bright-edged plateaus of similar-looking mare material. One interpretation is that these low plateaus are at the height of the original mare surface, but as the lavas cooled they deflated and collapsed. "Deflated" is a word first used for lava flows in Hawaii that were puffed up with gas, but as the flows cooled and the gas escaped, parts of the flow anchored by tree trunks remained at their original height. Perhaps the island at bottom left was anchored by the mountain ridge inside it.

Most young craters 40-km in diameter are classic complex craters, about 3-km deep with terraced walls and a central peak rising from a small floor. Kopff, within the Orientale Basin, sure is different. It has steep interior walls with no terraces, nor a central peak. The floor is very wide – about 90% of the crater's diameter – and covered by smooth material. These characteristics made some scientists question if Kopff was actually a volcanic caldera, but studies show that its walls are not made of volcanic material. However, the floor certainly is. This suggests that lavas welled up through the crust and erupted onto Kopff's original floor. The network of rilles that cut the floor look like expansion cracks that sometimes appear on the crust of baked bread; apparently after the floor lavas cooled, more volcanic material pushed up under the floor of Kopff, but rather than erupting, the it domed and cracked the floor.

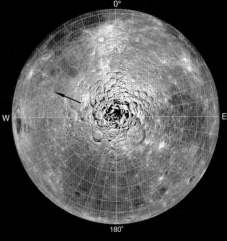

78: Inghirami

Location, location, location. The perfectly normal crater Inghirami had the misfortune to be near the edge of the massive deposits surging out from the formation of the Orien-tale Basin. Other preexisting craters that were closer were completely buried by vast surges of debris that raced over the landscape away from Orientale. One surge of material flowed from the left, over Inghirami's rim and terraces and reached partway across the floor before curving to the left around a central high zone. Amazing! Another river of ejecta, called the Inghirami Valley, flowed past the opposite side of the crater. You would not want to be anywhere near here when this happened about 3.8 billion years ago!

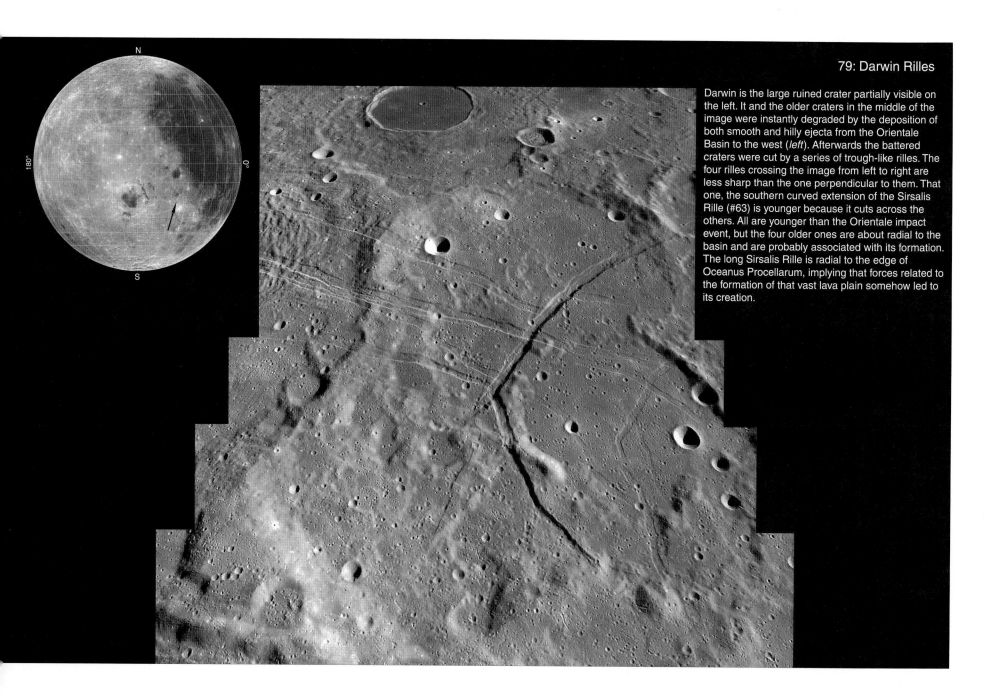

Darwin is the large ruined crater partially visible on the left. It and the older craters in the middle of the image were instantly degraded by the deposition of both smooth and hilly ejecta from the Orientale Basin to the west (*left*). Afterwards the battered craters were cut by a series of trough-like rilles. The four rilles crossing the image from left to right are less sharp than the one perpendicular to them. That one, the southern curved extension of the Sirsalis Rille (#63) is younger because it cuts across the others. All are younger than the Orientale impact event, but the four older ones are about radial to the basin and are probably associated with its formation. The long Sirsalis Rille is radial to the edge of Oceanus Procellarum, implying that forces related to the formation of that vast lava plain somehow led to its creation.

80: Leuschner Chain

Named after the crater partly visible at bottom right, the Leuschner Chain is the best example of a chain of secondary craters on the Moon. Its members are so large – typically 15–20 km wide – because they are secondaries from the formation of the great Orientale Basin. The basin is centered on the limb of the Moon as seen from Earth, and although its formation sent ejecta in all directions the debris was apprently thicker and most devastating on the western and southern sides. Nearly all of the terrain here is ejecta that is so thick that preexisting craters are completely buried. And this is 350 km from the basin!

81: Shackleton

Deep in the rugged southern highlands of the Moon is a simple 19-km-wide crater, Shackleton, whose left rim marks the south pole. This is of interest not just to mapmakers but more importantly for planners of future bases on the Moon. Because the Moon rotates so slowly – its day is about 29 Earth days – solar power is only available about half of the time anywhere on the Moon, except at the pole, where the Sun is low in the sky but nearly always visible. The poles are the only places on the Moon capable of generating continuous solar power; anyplace else will require nuclear reactors for uninterrupted energy. Shadow-filled Shackleton and other nearby polar craters offer another valuable resource. Because sunlight never shines onto the floors of some polar craters they contain ice deposited from billions of years of watery, cometary impacts. Water is necessary for humans to drink and can be broken into hydrogen and oxygen and used as rocket fuel. Shackleton may become the Saudi Arabia of the Moon.

82: Apollo Basin

Looking like a low spot puddled with lava, the Apollo impact basin on the Moon's farside is an older version of the more familiar Nectaris Basin visible from Earth. Both basins are small – Apollo is about 480 km across – and are nearly dry; lava floods only the center of the basins. Japanese scientists using *Kaguya* data discovered that lavas on Apollo's floor erupted at two different times, 2.4 and 3.5 billion years ago. The younger eruption emplaced the dark volcanics at the upper left of the lava-covered floor. The name Apollo was originally given to commemorate the *Apollo* missions to the Moon but has also become a memorial to America's fallen Challenger astronauts. All seven of the Challenger astronauts have craters named for them in Apollo, but only five are visible here.

Smith

Scobee

McAuliffe

Resnik

Onizuka

Jackson is the farside's Tycho (#44). It is young, has a magnificent ray system, and is surrounded by a dark collar of impact melt. And like Tycho it was formed by an oblique impact, with the projectile coming from the direction of horizon visible here and splashing impact melt downrange – that is, over the near rim and across the foreground. Jackson's rim is very unusual. The uprange side has normal terraces, but the nearside has a broad bench below the rim crest, which seems to be smeared with the same veneer of melt as beyond the rim. The floor repeats this uprange – downrange difference, with debris from the crater wall filling the farside of the floor, but being mostly absent from the nearside floor. A lot remains to be learned about oblique impacts!

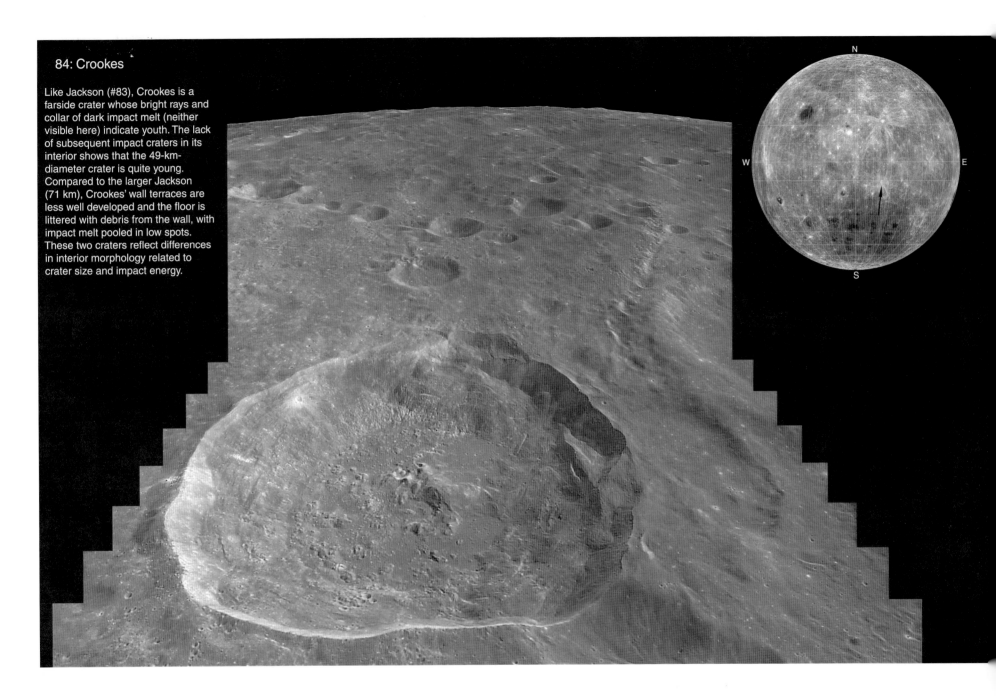

84: Crookes

Like Jackson (#83), Crookes is a farside crater whose bright rays and collar of dark impact melt (neither visible here) indicate youth. The lack of subsequent impact craters in its interior shows that the 49-km-diameter crater is quite young. Compared to the larger Jackson (71 km), Crookes' wall terraces are less well developed and the floor is littered with debris from the wall, with impact melt pooled in low spots. These two craters reflect differences in interior morphology related to crater size and impact energy.

Everyone knows that the highest place on Earth is Mt. Everest in the Himalaya Mountains. But until recently no one knew the location of the Moon's equivalent high point. The *Kaguya* spacecraft carried a laser altimeter that pinged the distance between the orbiter and the lunar surface more than 6 million times, and those data were used to create the most detailed topographic map of the Moon. Much of the lunar farside is higher than the Moon's average elevation, and the highest point is an unimpressive ridge (*arrow*) just beyond the eastern rim of the flat-floored crater Engelhardt. This point is 10,750 m above the average lunar elevation – there is no sea level to compare to. Mt Everest is quite a bit lower – 8,850 m above Earth's sea level – but much more dramatic!

86: Icarus

The crater Icarus was one of the first on the farside to be singled out. During the *Apollo 11* mission, when astronauts orbited the Moon, they noticed a crater nearly completely in shadow except for its central peak, which stuck up into sunlight. This is rare, for nearly all lunar craters have central peaks that are lower than their rims. In this view the Sun is high and the entire interior of Icarus is well illuminated. The crater seems to be a normal complex one, with terraces on the inner walls and a central peak. But the central peak looks more massive than normal and in fact it rises at least 4.1 km above the surrounding floor. Since the crater is about 4.6 km deep, the peak apparently is not higher than the rim but will be visible long before other parts of its interior become shadowed. Remarkable!

Because the Moon orbits Earth in the same amount of time that it rotates on its axis – its year and day are of the same duration– one side always faces Earth and the other side is always hidden to us. The lunar coordinate system is tied to Earth with the 0° of latitude and longitude crossing in a small maria called Sinus Medii or Central Bay on the Earth-facing side. Exactly on the opposite side of the Moon the 180° longitude meridian crosses 0° latitude at a totally indistinct point, here marked by the arrow. However, lunar pioneers probably will not use this point to navigate, instead referring to easily seen landmarks such as the 93-km-wide crater Daedalus in the foreground.

The Kaguya Lunar Atlas

88: Buys-Ballot

This is one of the strangest craters on the Moon. It is not circular but pear shaped, being about 90 km long by 60 km wide. Even more unusual is the elongated ridge in place of a central peak. Based on small-scale hypervelocity impact experiments in a laboratory, scientists know how this feature formed. For impacts that were vertical or even 60° away from vertical, the resulting craters were nearly circular. But when the impacting projectile came in at an angle of only 10° above the horizon, elongated craters with central ridges were produced. The existence of smooth dark material within Buys-Ballot and just outside its rim suggests that impact melt created by the crater formation was sloshed over the rim and deposited downrange. Buys-Ballot is a smaller version of Schiller (#71), which seems most likely also to have resulted from an oblique impact.

The morphology of impact craters changes with crater diameter. Craters smaller than about 15 km have simple bowl shapes, with walls that lead to small, flat floors. At larger diameters, craters become complex, with slumps and terraces on their walls and central peaks on their broad floors. At diameters larger than about 300 km craters are transformed into multi-ring impact basins such as Orientale. The transition from complex crater to basin is marked by only a few rare examples that have central peaks and poorly developed inner rings. Antoniadi and Schrödinger (#93) are the best lunar examples. Antoniadi has a diameter of 140 km, and its rim, outer portion of floor, and central peak look like a normal complex crater, but the discontinuous ring of mountains and hills, and the lavas inside the ring, mark the beginning of the transition toward multi-ring basin formation. The mechanism for this transformation is not well understood.

90: Mare Moscoviense

In 1959 the Soviets stunned the world when their spacecraft *Luna 3* photographed the always unseen farside of the Moon. The image was made during full Moon for the farside, and the image quality was terrible, but one dark feature stood out strikingly against the general brightness. The Soviets were so pleased with their great achievement that they named it Mare Moscoviense – the Sea of Moscow. Since nearly all other maria are named for often stormy weather conditions or states of mind, some wags wondered what the Russians meant. The mare is one of the few in the bright lunar highlands of the farside, filling the center of a multi-ring impact basin. Similar to the Apollo Basin, Moscoviense has become the place to honor fallen cosmonauts – Titov is the large crater in the mare and Komarov is crossed by fractures nearer the horizon.

91 Mare Ingenii

In the southern hemisphere of the farside is one of the most bizarre sights on the Moon. Mare lavas fill the 560-km-wide Ingenii impact basin, but what is odd is its ghostly pattern of curved ribbons of brightness. Like Reiner Gamma [#62] on the nearside, these swirls look like stains on the surface. They have no topographic relief, nor any associated small volcanoes or other features. The Ingenii swirls are associated with strong magnetic fields and are on the exact opposite side of the Moon from the Serenitatis Basin. The fact that some lunar swirls are antipodal to impact basins implies that a concentration of energy from major impacts causes swirls, but no one knows exactly how.

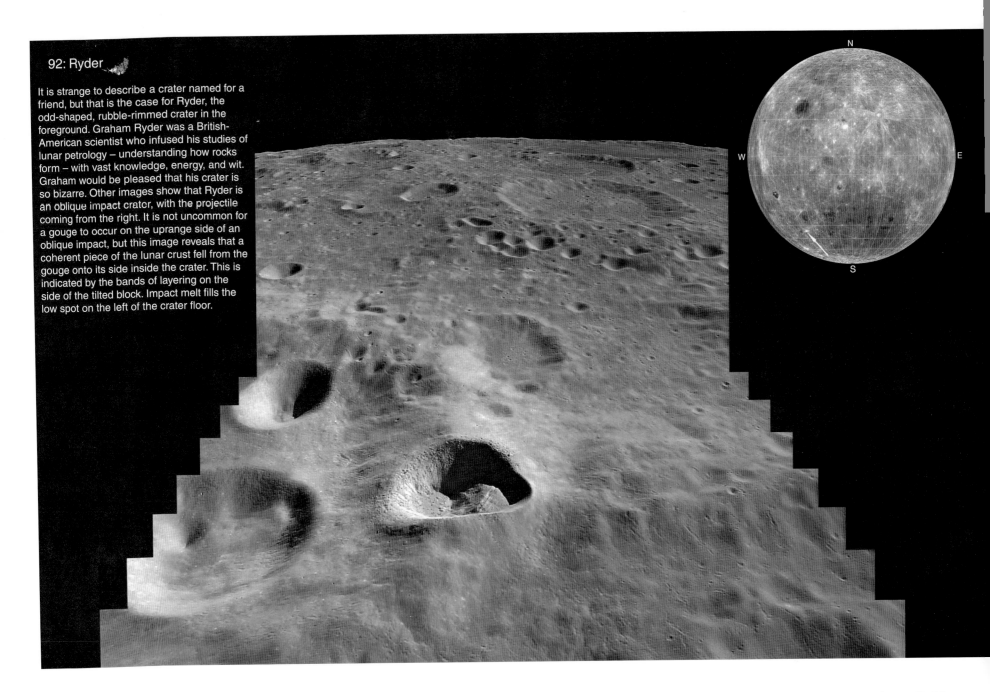

92: Ryder

It is strange to describe a crater named for a friend, but that is the case for Ryder, the odd-shaped, rubble-rimmed crater in the foreground. Graham Ryder was a British-American scientist who infused his studies of lunar petrology – understanding how rocks form – with vast knowledge, energy, and wit. Graham would be pleased that his crater is so bizarre. Other images show that Ryder is an oblique impact crater, with the projectile coming from the right. It is not uncommon for a gouge to occur on the uprange side of an oblique impact, but this image reveals that a coherent piece of the lunar crust fell from the gouge onto its side inside the crater. This is indicated by the bands of layering on the side of the tilted block. Impact melt fills the low spot on the left of the crater floor.

93: Schrödinger

Like Antoniadi (#89), Schrödinger is a transition structure, spanning the morphological divide between large, complex craters and true multi-ring impact basins. But with its diameter of 320 km, Schrödinger is nearly twice as large as Antoniadi and goes the next step, having no central peak, just a well-developed inner ring. At larger diameters, additional inner and outer rings develop around basins, giving a bull's-eye appearance. The moat of Schrödinger's floor – the area between the two rings – is part hilly and part smooth and is interpreted as a broad sheet of impact melt. The area within the inner ring is covered by apparent lavas with mare ridges, rilles, and a dark halo crater. Few subsequent impact craters were formed on Schrödinger, suggesting that it is younger than most basins but still about 3.9 billion years old.

94: Schrödinger Valley

Ejecta thrown out from impacts create a fountain-like scourge of debris falling in all directions. But sometimes clots of material appear to break apart during ejection, dropping piece by piece and creating a line of secondary craters. Around normal impact craters these crater chains are short and small. But when basins are formed they excavate and eject mountain-size clots of material that fall across the landscape, making long, continuous lines of secondary craters. This 310-km long line of overlapping secondaries radiates from the 320-km diameter Schrödinger Basin (#93). Where the chain crosses the older crater Sikorsky it climbs over the rim and drops down to the floor, clearly a deposit from above rather than a fracturing from below.

Most maria fill significant impact basins and are continuous oceans of solidified lava within well-defined rims. Mare Australe is an anomaly. It is an area about 900 km across that is defined by lava-filled craters such as those near the top. Australe is thought to be an ancient impact basin that could not hold onto its rim structures but whose deep fractures allowed younger mare lavas to rise to the surface and fill low spots. A remarkable occurrence is the continuation of a mare ridge across the highlands. The right side of the dark-floored crater Kugler (*top center*) contains a typical mare ridge, detectable even in the crater above it. Between Kugler and the large crater Anuchin (*bottom right*), the ridge continues across the highlands with a distinct edge, and unlike any other place on the Moon, the ridge covers the right half of a crater just above Anuchin. This ridge may have formed recently by the cooling of the Moon, slowly reducing its diameter.

0°

W   E

180°

96: Lacus Solitudinus

Just around the limb on the farside of the Moon seen only by orbiting spacecraft is this small outpost of mare volcanism. It is isolated from other volcanic regions; perhaps this is why it was named the Lake of Solitude. The most remarkable landform is the bathtub ring in the crater Bowditch near the bottom of the image. A single ledge 2–3 km wide edges the crater floor. In Hawaii, similar terraces are sometimes seen where the lava has lost the gas contained within it and subsided, leaving a ring around the edges where it cooled and becoming anchored. Considering that the Bowditch ledge is roughly 100 m above the rest of the floor, and that the crater is 40 km wide, a huge amount of gas must have been lost.

Like Ryder (#92), Necho is a young oblique impact crater on the farside. It has a broad terrace on the western (*left*) side, and the low side of the rim is opposite. The faint dark material beyond the rim on the right is impact melt sloshed out of the crater, perhaps by the collapse that formed the western mega-terrace. Most of Necho's rim is quite steep; the only ground route future astronauts can use to enter the crater is down the terraces and slump blocks on the west. Why would they want to go to the floor? To collect samples of the impact melt on the floor to determine the crater's date of formation and to see if any of the hills near the middle are central peaks. Because such peaks bounce up from depth, their analysis will help us understand the compositional layering of the lunar crust.

98: King

King is one of the most spectacular craters on the Moon. It is a relatively young crater, with a sharp-edged rim crest and terraces. Its central peak complex is connected to the crater wall by a debris slide, and on its opposite side – facing the camera – are two mountains lobes, all giving the appearance of a giant lobster claw. Just north of King (in front) is an older depression that is filled with a pool of smooth material – impact melt. The low side of King's rim is on this side. King resulted from yet another oblique impact, and it is likely that the wall collapse on the far side of the crater sloshed the impact melt over the low rim and into the depression. What a sight that would have been!

One of the first features discovered on the farside was the dark-floored crater Tsiolkovskiy. It is a typical large impact crater with terraced walls and a massive central peak. But 185–km wide Tsiolkovskiy is rare in having a floor of dark mare basalt. The farside crust is thicker than the nearside crust, probably accounting for the restriction of most farside mare to within the South Pole–Aitken Basin, which may have actually stripped the crust away. But Tsiolkovskiy is far the that basin; it is not known how magma happened to reach the surface there and very few other places on the farside.

100: Giordano Bruno

Lunar craters are named for scientists and philosophers, but this is the only one named for someone who was burned at the stake. Bruno was a Dominican priest of the late 1500s who supported Copernicanism and speculated that the universe was infinite and widely populated with intelligent beings. He was condemned as a heretic and executed by the Inquisition in 1600. The brightness of this farside crater and its extensive rays indicate that it is a young impact crater. It has been proposed that the crater actually formed in A. D. 1178, when English monks observed lights on the Moon at a position consistent with the location of Bruno. However, it is thought that ejecta from any significant lunar impact would cause a spectacular rain of shooting stars on Earth, and since that was not observed, most scientists doubt that the crater Giordano Bruno formed at that time. *Kaguya* scientists have estimated the age as 1–10 million years, very young by lunar standards, but too ancient to be witnessed by the monks!

Thumbnail Index

The following five pages depict each plate in the book and provide the following information about it:

- Longitude and latitude of the main feature shown.
- Sun's angle (SE), ranging from 1°, with grazing illumination and long shadows, up to 72° for nearly full Moon conditions with the Sun almost overhead.
- The elevation or height (H) in kilometers of the spacecraft above the surface when the image was acquired, from 21 to 116 km.
- The time of acquisition in this sequence: year, month, day, hour, and minute in Universal Time.

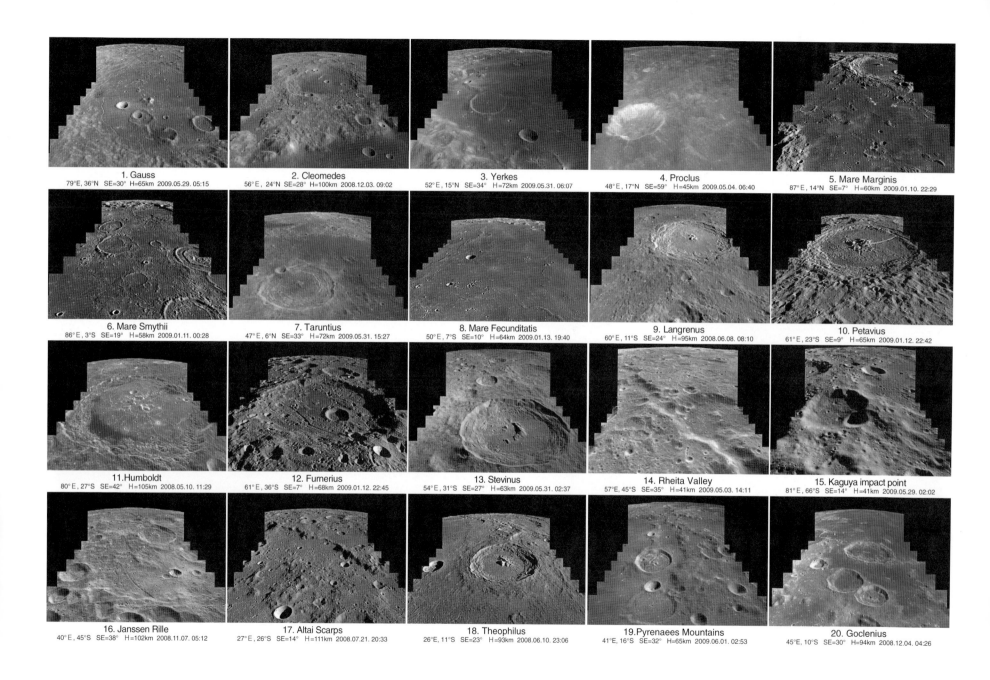

1. Gauss
79°E, 36°N SE=30° H=65km 2009.05.29. 05:15

2. Cleomedes
56°E, 24°N SE=28° H=100km 2008.12.03. 09:02

3. Yerkes
52°E, 15°N SE=34° H=72km 2009.05.31. 06:07

4. Proclus
48°E, 17°N SE=59° H=45km 2009.05.04. 06:40

5. Mare Marginis
87°E, 14°N SE=7° H=60km 2009.01.10. 22:29

6. Mare Smythii
86°E, 3°S SE=19° H=58km 2009.01.11. 00:28

7. Taruntius
47°E, 6°N SE=33° H=72km 2009.05.31. 15:27

8. Mare Fecunditatis
50°E, 7°S SE=10° H=64km 2009.01.13. 19:40

9. Langrenus
60°E, 11°S SE=24° H=95km 2008.06.08. 08:10

10. Petavius
61°E, 23°S SE=9° H=65km 2009.01.12. 22:42

11. Humboldt
80°E, 27°S SE=42° H=105km 2008.05.10. 11:29

12. Furnerius
61°E, 36°S SE=7° H=68km 2009.01.12. 22:45

13. Stevinus
54°E, 31°S SE=27° H=63km 2009.05.31. 02:37

14. Rheita Valley
57°E, 45°S SE=35° H=41km 2009.05.03. 14:11

15. Kaguya impact point
81°E, 66°S SE=14° H=41km 2009.05.29. 02:02

16. Janssen Rille
40°E, 45°S SE=38° H=102km 2008.11.07. 05:12

17. Altai Scarps
27°E, 26°S SE=14° H=111km 2008.07.21. 20:33

18. Theophilus
26°E, 11°S SE=23° H=93km 2008.06.10. 23:06

19. Pyrenaees Mountains
41°E, 16°S SE=32° H=65km 2009.06.01. 02:53

20. Goclenius
45°E, 10°S SE=30° H=94km 2008.12.04. 04:26

The Kaguya Lunar Atlas
164

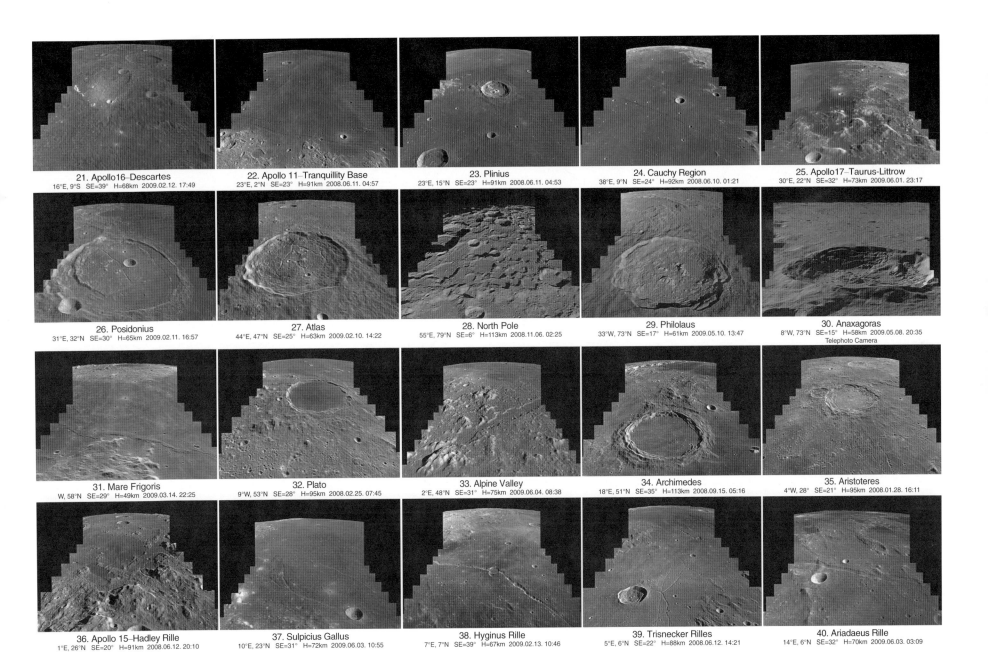

21. Apollo16–Descartes
16°E, 9°S SE=39° H=68km 2009.02.12. 17:49

22. Apollo 11–Tranquillity Base
23°E, 2°N SE=23° H=91km 2008.06.11. 04:57

23. Plinius
23°E, 15°N SE=23° H=91km 2008.06.11. 04:53

24. Cauchy Region
38°E, 9°N SE=24° H=92km 2008.06.10. 01:21

25. Apollo17–Taurus-Littrow
30°E, 22°N SE=32° H=73km 2009.06.01. 23:17

26. Posidonius
31°E, 32°N SE=30° H=65km 2009.02.11. 16:57

27. Atlas
44°E, 47°N SE=25° H=63km 2009.02.10. 14:22

28. North Pole
55°E, 79°N SE=6° H=113km 2008.11.06. 02:25

29. Philolaus
33°W, 73°N SE=17° H=61km 2009.05.10. 13:47

30. Anaxagoras
8°W, 73°N SE=15° H=58km 2009.05.08. 20:35
Telephoto Camera

31. Mare Frigoris
W, 58°N SE=29° H=49km 2009.03.14. 22:25

32. Plato
9°W, 53°N SE=28° H=95km 2008.02.25. 07:45

33. Alpine Valley
2°E, 48°N SE=31° H=75km 2009.06.04. 08:38

34. Archimedes
18°E, 51°N SE=35° H=113km 2008.09.15. 05:16

35. Aristoteres
4°W, 28° SE=21° H=95km 2008.01.28. 16:11

36. Apollo 15–Hadley Rille
1°E, 26°N SE=20° H=91km 2008.06.12. 20:10

37. Sulpicius Gallus
10°E, 23°N SE=31° H=72km 2009.06.03. 10:55

38. Hyginus Rille
7°E, 7°N SE=39° H=67km 2009.02.13. 10:46

39. Trisnecker Rilles
5°E, 6°N SE=22° H=88km 2008.06.12. 14:21

40. Ariadaeus Rille
14°E, 6°N SE=32° H=70km 2009.06.03. 03:09

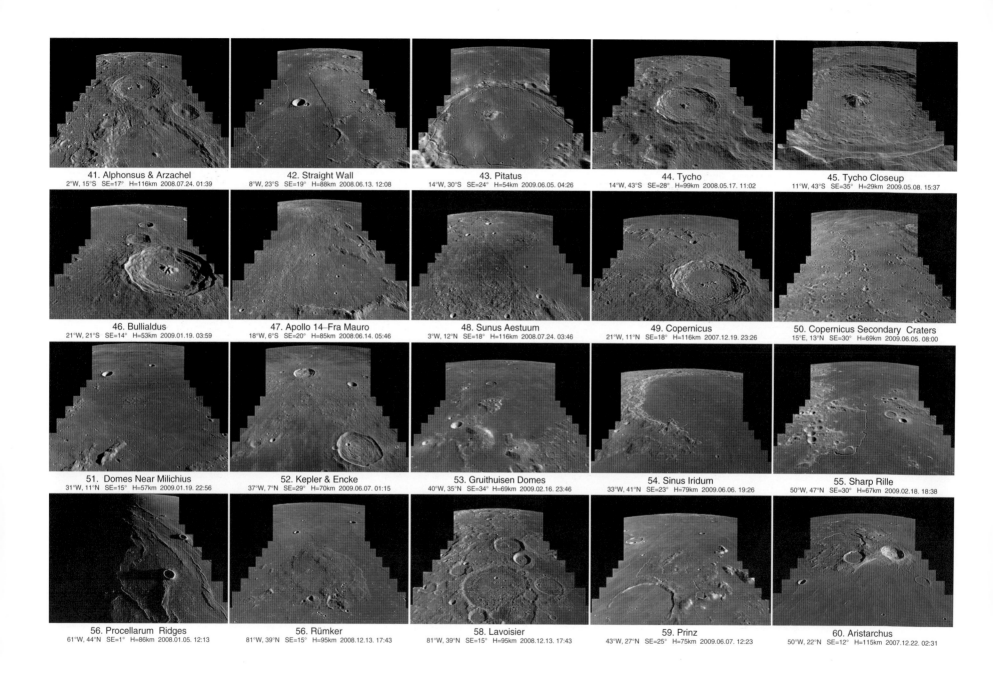

41. Alphonsus & Arzachel
2°W, 15°S SE=17° H=116km 2008.07.24. 01:39

42. Straight Wall
8°W, 23°S SE=19° H=88km 2008.06.13. 12:08

43. Pitatus
14°W, 30°S SE=24° H=54km 2009.06.05. 04:26

44. Tycho
14°W, 43°S SE=28° H=99km 2008.05.17. 11:02

45. Tycho Closeup
11°W, 43°S SE=35° H=29km 2009.05.08. 15:37

46. Bullialdus
21°W, 21°S SE=14° H=53km 2009.01.19. 03:59

47. Apollo 14–Fra Mauro
18°W, 6°S SE=20° H=85km 2008.06.14. 05:46

48. Sunus Aestuum
3°W, 12°N SE=18° H=116km 2008.07.24. 03:46

49. Copernicus
21°W, 11°N SE=18° H=116km 2007.12.19. 23:26

50. Copernicus Secondary Craters
15°E, 13°N SE=30° H=69km 2009.06.05. 08:00

51. Domes Near Milichius
31°W, 11°N SE=15° H=57km 2009.01.19. 22:56

52. Kepler & Encke
37°W, 7°N SE=29° H=70km 2009.06.07. 01:15

53. Gruithuisen Domes
40°W, 35°N SE=34° H=69km 2009.02.16. 23:46

54. Sinus Iridum
33°W, 41°N SE=23° H=79km 2009.06.06. 19:26

55. Sharp Rille
50°W, 47°N SE=30° H=67km 2009.02.18. 18:38

56. Procellarum Ridges
61°W, 44°N SE=1° H=86km 2008.01.05. 12:13

56. Rümker
81°W, 39°N SE=15° H=95km 2008.12.13. 17:43

58. Lavoisier
81°W, 39°N SE=15° H=95km 2008.12.13. 17:43

59. Prinz
43°W, 27°N SE=25° H=75km 2009.06.07. 12:23

60. Aristarchus
50°W, 22°N SE=12° H=115km 2007.12.22. 02:31

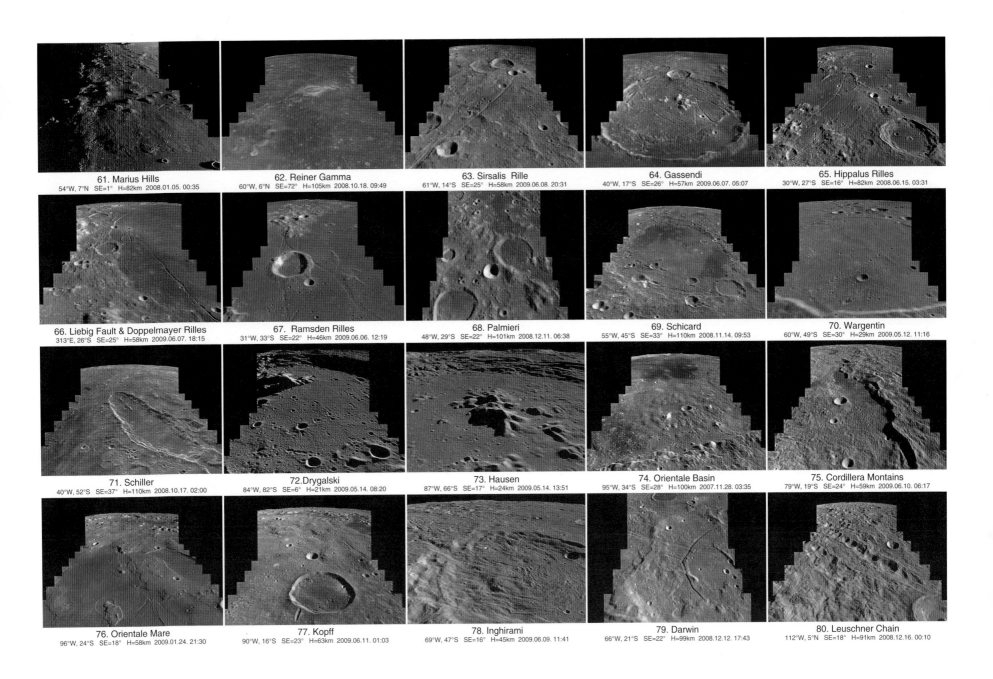

61. Marius Hills
54°W, 7°N SE=1° H=82km 2008.01.05. 00:35

62. Reiner Gamma
60°W, 6°N SE=72° H=105km 2008.10.18. 09:49

63. Sirsalis Rille
61°W, 14°S SE=25° H=58km 2009.06.08. 20:31

64. Gassendi
40°W, 17°S SE=26° H=57km 2009.06.07. 05:07

65. Hippalus Rilles
30°W, 27°S SE=16° H=82km 2008.06.15. 03:31

66. Liebig Fault & Doppelmayer Rilles
313°E, 26°S SE=25° H=58km 2009.06.07. 18:15

67. Ramsden Rilles
31°W, 33°S SE=22° H=46km 2009.06.06. 12:19

68. Palmieri
48°W, 29°S SE=22° H=101km 2008.12.11. 06:38

69. Schicard
55°W, 45°S SE=33° H=110km 2008.11.14. 09:53

70. Wargentin
60°W, 49°S SE=30° H=29km 2009.05.12. 11:16

71. Schiller
40°W, 52°S SE=37° H=110km 2008.10.17. 02:00

72.Drygalski
84°W, 82°S SE=6° H=21km 2009.05.14. 08:20

73. Hausen
87°W, 66°S SE=17° H=24km 2009.05.14. 13:51

74. Orientale Basin
95°W, 34°S SE=28° H=100km 2007.11.28. 03:35

75. Cordillera Montains
79°W, 19°S SE=24° H=59km 2009.06.10. 06:17

76. Orientale Mare
96°W, 24°S SE=18° H=58km 2009.01.24. 21:30

77. Kopff
90°W, 16°S SE=23° H=63km 2009.06.11. 01:03

78. Inghirami
69°W, 47°S SE=16° H=45km 2009.06.09. 11:41

79. Darwin
66°W, 21°S SE=22° H=99km 2008.12.12. 17:43

80. Leuschner Chain
112°W, 5°N SE=18° H=91km 2008.12.16. 00:10

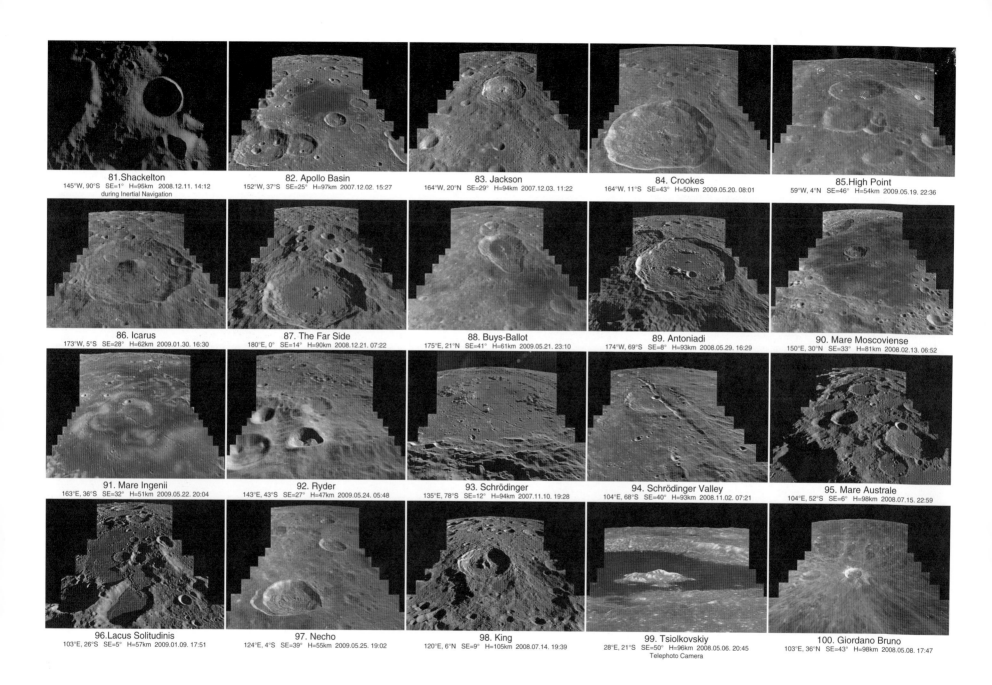

81.Shackelton
145°W, 90°S   SE=1°   H=95km   2008.12.11. 14:12
during Inertial Navigation

82. Apollo Basin
152°W, 37°S   SE=25°   H=97km   2007.12.02. 15:27

83. Jackson
164°W, 20°N   SE=29°   H=94km   2007.12.03. 11:22

84. Crookes
164°W, 11°S   SE=43°   H=50km   2009.05.20. 08:01

85.High Point
59°W, 4°N   SE=46°   H=54km   2009.05.19. 22:36

86. Icarus
173°W, 5°S   SE=28°   H=62km   2009.01.30. 16:30

87. The Far Side
180°E, 0°   SE=14°   H=90km   2008.12.21. 07:22

88. Buys-Ballot
175°E, 21°N   SE=41°   H=61km   2009.05.21. 23:10

89. Antoniadi
174°W, 69°S   SE=8°   H=93km   2008.05.29. 16:29

90. Mare Moscoviense
150°E, 30°N   SE=33°   H=81km   2008.02.13. 06:52

91. Mare Ingenii
163°E, 36°S   SE=32°   H=51km   2009.05.22. 20:04

92. Ryder
143°E, 43°S   SE=27°   H=47km   2009.05.24. 05:48

93. Schrödinger
135°E, 78°S   SE=12°   H=94km   2007.11.10. 19:28

94. Schrödinger Valley
104°E, 68°S   SE=40°   H=93km   2008.11.02. 07:21

95. Mare Australe
104°E, 52°S   SE=6°   H=98km   2008.07.15. 22:59

96.Lacus Solitudinis
103°E, 26°S   SE=5°   H=57km   2009.01.09. 17:51

97. Necho
124°E, 4°S   SE=39°   H=55km   2009.05.25. 19:02

98. King
120°E, 6°N   SE=9°   H=105km   2008.07.14. 19:39

99. Tsiolkovskiy
28°E, 21°S   SE=50°   H=96km   2008.05.06. 20:45
Telephoto Camera

100. Giordano Bruno
103°E, 36°N   SE=43°   H=98km   2008.05.08. 17:47

Subject Index

Below are terms and named features shown on the atlas plates. The numbers refer to plate numbers, not page numbers.

A

age 44, 100
Aldrin 22
Alpetragius 41
Alphonsus 41
Alpine Valley 33
Alps 33
Altai Scarp 17, 75
Anaxagoras 30
Antoniadi 89
Anuchin 95
Apennine Mountains 3
Apollo 11 22
Apollo 14 47
Apollo 15 36
Apollo 16 21
Apollo 17 25
Apollo Basin 82
Archimedes 35
Ariadaeus Rille 40
Aristarchus 60
Aristarchus Plateau 60
Aristoteles 34
Armstrong 22
Arzachel 41
ash 1, 2, 48, 60, 61
astronaut 82
Atlas 27

B

basin 93
basin ejecta 28, 31, 47, 75, 78, 79, 80
basin secondary crater 80
basin secondary crater chain 14, 94
Birt 42
Bode II Rille 48
Bohnenberger 19
Bowditch 96
Bullialdus 46
Buys-Ballot 88

C

caldera 38
Campanus 65
Catena Leuschner 80
Cauchy 24
Cauchy Fault 24
Cauchy Rille 24
central peak 35, 49, 86, 97
central peak basin 89
Challenger astronauts 82
Cleomedes 2
collapse pit 38
Collins 22
complex crater 49
concentric crater 58

cooling crack 45
coordinate system 87
Copernicus 35, 49, 50
Cordillera Mountains 74, 75
cosmonaut 90
Crookes 84
crust thickness 99

D
Daedalus 87
dark halo crater 1, 2, 27, 41, 69, 70, 93
Darwin 79
Descartes 21
dike 40, 63, 67
Dolland 21
dome 51, 57
Doppelmayer Rilles 66
Drygalski 72

E
ejecta 4, 16, 44, 45, 69
ejecta surge 78
Encke 52
Engelhardt 85
erosion 44

F
Fabricius 16
farside 87
fault 41, 42, 54, 66
fire fountain 48
floor-fractured crater 1, 6, 7, 10, 11, 19, 20, 27, 41, 43, 52, 58, 64
Fra Mauro 47
Furnerius 12

G
Gassendi 64
Gauss 1
ghost crater 8

Gill 15
Giordano Bruno 100
glass bead 25
Goclenius 20
Goclenius Rille 20
graben 63
Gruithuisen 53
Gruithuisen Domes 53

H
Hadley Rille 36
Haldane 6
Harbinger Mountains 59
Hausen 73
Hawaii 38
Heraclides Promontory 54
high point 85
highland ridge 95
Hippalus 65
history 15, 36, 90
Humboldt 11
Humorum basin 67, 68
Hyginus 38
Hyginus Rille 38, 39

I
Ibn Battuta 8
Icarus 86
Imbrium Basin 28, 31, 33, 39, 54
Imbrium Basin ejecta 52
impact debris 21
impact melt 9, 13, 18, 34, 45, 49, 74, 83, 84, 88, 92, 93, 97, 98
Inghirami 78
Inghirami Valley 78

J
Jackson 83
Janssen 16
Janssen Rille 16

K

Kaguya impact point 15
Kepler 52
Kies 46
King 98
Komarov 90
Kopff 77
Kugler 95

L

Lacus Solitudinus 96
Langrenus 9, 13
lava composition 53
lava filling 35
Lavoisier 58
Leuschner 80
Leuschner Chain 80
Lick 3
Liebig Fault 66
linear rille 40
Lohse 9
Luna 2 36
Luna 3 90

M

magnetic field 5, 62
magnetism 91
Mallet 14
Mare Australe 95
Mare Crisium 2, 3
Mare Fecunditatis 8, 20
Mare Frigoris 31
Mare Humorum 65, 66
Mare Imbrium 32, 50, 54
Mare Ingenii 91
Mare Insularum 51
mare lavas 69
Mare Marginis 5
Mare Moscoviense 90
Mare Nubium 42, 43, 46

Mare Orientale 76
mare ridge 3, 31, 56, 93, 95
Mare Serenitatis 23, 26, 37
Mare Smythii 6
Mare Tranquillitatis 22, 23
Mare Vaporum 31
Marius Hills 61
Milichius 51
Milichius Pi dome 51
Mitchell 34
Montes Alpes 33
Montes Apenninus 3
Montes Cordillera 74, 75
Montes Harbinger 59
Montes Pyrenaeus 19
Montes Rook 74
moonquake 29

N

Necho 97
Nectaris Basin 12, 14, 16, 17,
75, 82
Neper 5
north pole 28

O

oblique impact 4, 71, 83, 88, 92, 97
Oceanus Procellarum 52, 55, 56,
57, 58, 59, 63, 79
Orientale Basin 69, 74, 75, 76, 77, 79, 80
Orientale Basin ejecta 70, 72, 78

P

Palmieri 68
Petavius 10
Philolaus 29
Pitatus 43
Plato 32, 72
Plinius 23
Posidonius 26

Prinz 59
Proclus 4
Promontorium Heraclides 54
Pyrenees Mountains 19
pyroclastics 25, 37, 48, 60, 66

R
Ramsden 67
Ramsden Rilles 67
ray 4, 43
Reiner Gamma 62, 91
Rheita Valley 14
rille 1, 2, 10, 12, 20, 26, 27, 37, 39, 41,
42, 43, 48, 51, 55, 63, 64, 65, 66,
67, 68, 77, 79, 93
Rima Ariadaeus 40
Rima Bode II 48
Rima Cauchy 24
Rima Hadley 36
Rima Hyginus 38, 39
Rima Janssen 16
Rimae Doppermayer 66
Rimae Goclenius 20
Rimae Ramdsen 67
Rook Mountains 74
Ross 23
Rümker 57
Rupes Altai 17, 75
Rupes Cauchy 24
Rupes Liebig 66
Rupes Recta 42
Ryder 92

S
scallop 29
Schickard 69
Schiller 71, 88
Schrödinger 93
Schrödinger Valley 94
Schröter's Valley 60
secondary crater 43, 46, 50

Serenitatis basin 25
Shackleton 81
Sharp Rille 55
Sikorsky 94
sinuous rille 26, 55, 59
Sinus Aestuum 31, 48
Sinus Iridum 54
Sinus Medii 87
Sirsalis 63
Sirsalis Rille 63, 79
Snellius Valley 12
solar energy 28, 81
south pole 81
South Pole-Aitken Basin 99
South Ray Crater 21
spacecraft impact 15
Stevinus 13
Straight Wall 42
Sulpicius Gallus 37
swirl 5, 62, 91

T
Taruntius 7
Taurus-Littrow Valley 25
terrace 18, 30, 49
Theophilus 18, 72
Titov 90
topography 73, 85, 86
Tranquillity Base 22
transient lunar phenomena 32
trench 10
Triesnecker 39
Tsiolkovskiy 99
Tycho 43, 44, 45, 46, 83

V
Vallis Alpes 33
Vallis Inghirami 78
Vallis Rheita 14
Vallis Schrödinger 94
Vallis Schröteri 60

Vallis Snellius 12
volatiles 61
volcanic cone 61
von Braun 58

W
wall slump 23

Wargentin 70
water 28, 81
wrinkle ridge 56

Y
Yerkes 3

Printed in the United States of America